Studies in Systems, Decision and Control

Volume 473

Series Editor

Janusz Kacprzyk, Systems Research Institute, Polish Academy of Sciences, Warsaw, Poland

The series "Studies in Systems, Decision and Control" (SSDC) covers both new developments and advances, as well as the state of the art, in the various areas of broadly perceived systems, decision making and control–quickly, up to date and with a high quality. The intent is to cover the theory, applications, and perspectives on the state of the art and future developments relevant to systems, decision making, control, complex processes and related areas, as embedded in the fields of engineering, computer science, physics, economics, social and life sciences, as well as the paradigms and methodologies behind them. The series contains monographs, textbooks, lecture notes and edited volumes in systems, decision making and control spanning the areas of Cyber-Physical Systems, Autonomous Systems, Sensor Networks, Control Systems, Energy Systems, Automotive Systems, Biological Systems, Vehicular Networking and Connected Vehicles, Aerospace Systems, Automation, Manufacturing, Smart Grids, Nonlinear Systems, Power Systems, Robotics, Social Systems, Economic Systems and other. Of particular value to both the contributors and the readership are the short publication timeframe and the world-wide distribution and exposure which enable both a wide and rapid dissemination of research output.

Indexed by SCOPUS, DBLP, WTI Frankfurt eG, zbMATH, SCImago.

All books published in the series are submitted for consideration in Web of Science.

He Li · Weiwen Peng · Sidum Adumene ·
Mohammad Yazdi

Intelligent Reliability and Maintainability of Energy Infrastructure Assets

 Springer

He Li ⓘ
Centre for Marine Technology and Ocean
Engineering (CENTEC), Instituto
SuperiorTécnico
Universidade de Lisboa
Lisbon, Portugal

Sidum Adumene ⓘ
School of Ocean Technology, Fisheries
and Marine Institute
Memorial University of Newfoundland
St. John's, Canada

Weiwen Peng ⓘ
School of Intelligent Systems Engineering
Sun Yat-sen University
Shenzhen, China

Mohammad Yazdi ⓘ
Faculty of Science and Engineering
Macquarie University
Sydney, Australia

School of Computing, Engineering
and Physical Sciences
University of the West of Scotland
London, UK

ISSN 2198-4182   ISSN 2198-4190   (electronic)
Studies in Systems, Decision and Control
ISBN 978-3-031-29964-3   ISBN 978-3-031-29962-9   (eBook)
https://doi.org/10.1007/978-3-031-29962-9

To my dad Zhida, mom Yuqin

He Li

*To my dad Liwang, mom Qingmei, wife
Mingzhu and daughter Longyi*

Weiwen Peng

*To my wife—Barigiasi, and my lovely
children—Tegio, Giodu and Popka*

Sidum Adumene

To my lovely Rosita

Mohammad Yazdi

Preface

Intelligent reliability and maintainability of energy infrastructure assets refer to the use of advanced technologies and analytics to improve the reliability and maintenance of energy infrastructure, such as power plants, transmission networks, and renewable and non-renewable energy sources. This can be achieved through real-time monitoring and predictive analytics, condition-based maintenance, and automation of maintenance processes. The goal is to increase the efficiency and cost-effectiveness of maintaining energy infrastructure while ensuring that the assets operate at optimal levels and reducing the risk of outages and other disruptions.

This book provides valuable insight into practical and advanced decision-making methods under different mathematical models in energy infrastructure assets' intelligent reliability and maintainability. The explained case studies highlight the applicability of each proposed approach, which can be helpful for complex systems and empower the reliability and maintainability of energy infrastructure assets.

This book is organized to include nine chapters. Chapter 1 reviews the state-of-the-art research on the reliability and maintainability of energy infrastructure assets and emphasizes the intelligent tools and methods proposed from a bibliometric and literature review point of view. The purpose of this chapter is to show the practical need for the ideas and methodologies presented in this book.

Chapter 2 provides an overview of the energy infrastructures' system safety, reliability, maintainability, and resilience frameworks published in the past decades. This subject discusses frontier directions and development trends to reveal the research status. Finally, a bibliometric study is undertaken to ascertain the most productive and influential researchers, research centers, and hotspot fields.

Chapter 3 aims to study the application of a dynamic Bayesian modeling network to manage a critical offshore infrastructure under the influence of the material degradation process. The developed model is applied to burst probabilistic pressure modeling for safety improvement over time. The design methods suggested by ASME B31G and other empirical formulation methods are utilized to evaluate the burst pressure.

Chapter 4 introduces an improved LeNet-5 convolutional neural network model for condition-based maintenance and fault diagnosis of bearings. The model can effectively extract the fault features of one-dimensional vibration signals. Specifically, time–frequency transformation methods, short-time Fourier transform, and wavelet transform are used to convert the vibration signal into a two-dimensional time-domain–frequency-domain signal so as to extend the amount of time-domain–frequency-domain information of the vibration signal.

Chapter 5 introduces a Global Average Pooling-based Convolutional Siamese Network for the fault diagnosis of planetary gearboxes, which can cope with limited data situations. In the model, new ideas about feature extraction capability improvement, dimensionality reduction of features, and classification capability improvement are achieved.

Chapter 6 explores the various methodologies for failure assessment of subsea systems considering the unstable operating environment as well as functional dependencies among sub-components. The various failure influencing factors for harsh offshore operations were examined to establish their level of importance and impact on the failure model prediction for the subsea assets. The systematic study explores the failure methods that integrate the data-driven approaches with the physics of failure models to enhance better failure risk prediction in a dynamic offshore environment.

Chapter 7 proposes the application of an integrated probabilistic model for the failure consequence assessment of oil and gas pipelines suffering under-deposit corrosion (UDC). The physics of UDC potential based on the degradation mechanism is explored and built into a network to predict the corrosion rate. The predicted corrosion rate under the dominant corrosion mechanism is mapped into a probabilistic structure to capture their interactions on the failure state of the asset.

Chapter 8 investigates sustainable circular economy in the design and development of energy infrastructures. Thus, the risk management of a supercritical water gasification system as a critical energy infrastructure has been studied to identify the relevant implication in practice, gaps in the existing knowledge, and cutting-edge strategies to empower the circular economy in the lifecycle of energy infrastructures, such as recertification, refurbishment, remanufacturing, and recycling, to decommission, landfill, and more.

Chapter 9 considers field research, interviewing with the experts, to the best of our literature review, and utilization of artificial intelligence tools (e.g., Openai), we noticed the series of energy infrastructure challenges and how to deal with them. As a result, future studies in the energy infrastructure sector should focus on grid optimization, energy demand management, decentralized energy systems, financing and investment, energy transition and low-carbon development, rural electrification, energy policy and governance, and public awareness and engagement. These studies will provide valuable insights and guidance for the continued development of a robust and sustainable energy infrastructure.

This book will be one of the valuable guidance books for professionals and researchers working on system safety, reliability, and maintainability. It also aims to become a valuable reference book for postgraduate and undergraduate students.

Finally, as the authors of the present book, we are grateful to our family and friends for their constant love, patience, and support. You have all contributed through your words of encouragement for the period of the present work, with the best wishes that the book will be helpful to all concerned.

Lisbon, Portugal   Dr. He Li
Shenzhen, China   Dr. Weiwen Peng
St. John's, Canada   Dr. Sidum Adumene
Sydney, Australia   Dr. Mohammad Yazdi
February 2023

Contents

Chapter 1
Advances in Intelligent Reliability and Maintainability of Energy Infrastructure Assets

Abstract This chapter reviews the state-of-the-art research on the reliability and maintainability of energy infrastructure assets and emphasizes the Intelligent tools and methods proposed from a bibliometric and literature review point of view. The purpose of this chapter is to show the practical need for the ideas and methodologies presented in this book. A brief introduction to the reliability and maintainability of energy infrastructure assets is finalized. Subsequently, details of advances in reliability and the maintainability of energy infrastructure assets are summarized. The potential opportunities of the related research and engineering applications are also provided in this chapter.

Keywords Energy infrastructure assets · Reliability · Maintainability · Intelligent tools

1.1 Introduction

Energy, fossil-based and renewable, is the foundation of human society. It supports the development and progress of human beings [1]. Energy infrastructure assets include a selection of equipment and systems and their appendant items and structures directly and indirectly related to energy generation [2, 3].

Reliability and maintainability are critical performance indices of engineering equipment and systems. Reliability is a degree to which energy infrastructure assets can operate as designed under the given working condition and observation time [4–6]. It is also known as a capability of energy infrastructure assets working without any failure, and such a capability is described by probability. The reliability of energy infrastructure assets is formatted from the design, manufacturing, installation, operation, and maintenance process and is affected by multiple factors like human, machine, materials, tools (manufacturing measures), and environment [7]. Reliability is a required indicator of the design of systems, that is, the designed system must fulfill its reliability standard [8]. For instance, several wind energy projects stipulate that wind turbines should preserve a certain level of reliability at the end of their life span, usually 20–25 years.

© The Author(s), under exclusive license to Springer Nature Switzerland AG 2023
H. Li et al., *Intelligent Reliability and Maintainability of Energy Infrastructure Assets*,
Studies in Systems, Decision and Control 473,
https://doi.org/10.1007/978-3-031-29962-9_1

Reliability in energy infrastructure refers to the ability of energy systems to deliver a consistent and uninterrupted supply of electricity, heat, or fuel to end users. The reliability of energy infrastructure is crucial for the functioning of modern societies, as it supports critical activities such as transportation, manufacturing, healthcare, and communication. There are several factors affect the reliability of energy infrastructure, including the quality of equipment and materials used in energy production, transmission, and distribution; the design and layout of the infrastructure; the adequacy of maintenance and repair activities; and the robustness of backup systems and contingency plans. Reliability is associated with key performance factors of energy infrastructure assets and their subsystems, components, and essential parts, including, but not limited to, failure probability, failure rate, and mean time to failure [9–12]. Reliability investigations are to [12–17]: (i) find out failures that may happen to energy infrastructure assets and identify critical ones; (ii) find out preventive and corrective actions to prevent critical failures of energy infrastructure assets; (iii) infer reliability-related indices of energy infrastructure assets such as failure rate and mean time to failure to understand failure features of such devices; (iv) determine failure prevention and reliability improvement actions that may be applied to energy infrastructure assets to guarantee their robust performance during life span.

Maintainability in energy infrastructure refers to the efficiency with which energy systems can be maintained, repaired, and upgraded over their lifecycle. Maintaining the reliability and efficiency of energy infrastructure requires regular maintenance and periodic replacement or upgrading of equipment, systems, and components. Maintainability is important in energy infrastructure because it can affect the availability and operational efficiency of energy systems. Poor maintainability can result in increased downtime, decreased energy efficiency, and increased costs associated with maintenance and repair activities. To ensure good maintainability of energy infrastructure, several factors should be considered during the design and construction stages, including ease of access to equipment and systems, the use of standard components and materials, clear labeling and marking of equipment, and the provision of sufficient space for maintenance activities. Maintainability represents the capability of energy infrastructure assets that can be fixed on time. It includes maintenance strategy planning and associated issues like maintenance resources arrangement, spare part policy, and so on [18]. Take the floating offshore wind farm as an example, the maintainability of which includes, at least, maintenance strategy planning to decide the time to maintenance, maintenance resource arrangement to arrange maintenance crew and vessels in a cost-saving way, spare part dispatch and logistics to manage replaceable materials among manufactures, store rooms, and wind farms, and weather window determination to decide the allowable time to start maintenance [19]. Maintenance strategy planning is crucial among the maintainability above sectors of energy infrastructure assets as other sectors are usually operated and managed based on the determined maintenance strategy planning [20–22].

There are three basic maintenance strategies: preventive, corrective, and condition-based maintenance [22, 23]. To be specific, preventive planning is to inspect the health state of energy infrastructure assets periodically before unwanted failures happen. The inspection interval and the start threshold determination are the keys to preventive maintenance [24]. Corrective maintenance is passive maintenance that initializes the maintenance process of energy infrastructure assets after failures occur [25]. Condition-based maintenance is a relatively new concept that analyzes the health state of energy infrastructure assets with the monitored data from sensors [26]. It can make a real-time decision on the maintenance time of energy infrastructure assets. The core of condition-based maintenance is sensor data collection and analysis, as well as a decision-making model for maintenance threshold [27].

It is worth mentioning that predictive maintenance has emerged recently, which has become promising in energy infrastructure assets operation and maintenance [28]. Opportunistic maintenance is also proposed and applied to modern systems. It combines preventive and corrective maintenance and allows the maintenance crew to conduct additional inspections after corrective maintenance has been done [29]. Opportunistic maintenance is particularly suitable for energy infrastructure assets with low accessibility of maintenance and high logistic costs [30].

With the advances of material science, computer science, decision-making tools, and many other subjects, intelligent models, methodologies, and tools are developed and possible to be applied in reliability and maintainability-related issues of energy infrastructure assets [27, 31]. It provides engineers with an intelligent way of operating and managing today's energy infrastructure assets and is extendable to other complicated systems and equipment.

According to those above, one should realize that reliability and maintainability are critical performance factors of energy infrastructure assets throughout their life span. They are the basis of their operation and maintenance activities. Accordingly, this chapter reviews state-of-the-art research on the reliability and maintainability of energy infrastructure assets. The purpose of this chapter is to provide an overall framework of this book from an academic point of view.

The rest of this chapter is arranged as follows. Section 1.2 introduces the data collection and source of this literature review. Section 1.3 provides advances and trends in intelligent reliability investigation of energy infrastructure assets. Similar targets on the maintainability aspect are displayed in Sect. 1.4. Critical discussions are provided in Sect. 1.5. The authors conclude this chapter with Sect. 1.6.

1.2   Data Collection and Data Source

The data source is the Web of Science core collection. Specific data collection information is provided in Table 1.1. This chapter reviews the reliability and maintainability aspects of energy infrastructure assets so as to provide the readers with a clear framework of the related field. Specifically, two search topics are completed: (i) reliability of energy infrastructure assets; and (ii) maintainability of energy infrastructure assets.

The search is conducted as a subject search for articles published from 1922 to February 2023, 100 years in total. There were 587 articles under the maintainability of energy infrastructure assets theme and 560 articles under the reliability of energy

Table 1.1 Detailed information on literature searching

Database	Search method	Search terms	Time frame	Number of results	Time	Record content
Web of science core collection	Topic	Maintenance of energy asset	From the whole year 1992 to 2023	587	December 10th 2022	Full record and cited references
		Reliability of energy asset		560		

infrastructure assets theme. Retrieved on December 10th, 2022. The data file record format is Full Record and Cited References with TXT File. The software used for this keyword analysis is VOSviewer 1.6.18.

1.3   Advances in Intelligent Reliability Investigations of Energy Infrastructure Assets

1.3.1   Bibliometric

In this part, keywords are searched with the topic of the reliability of energy infrastructure assets. The minimum occurrence number of a keyword is set to 10 to exclude those that are not popular or interested by recent research. Overall, 34 out of the final 2539 keywords were filtered per the composite threshold, see Table 1.2. Table 1.2 is ordered by frequency of keyword occurrence, from highest to lowest.

According to Table 1.2, there are four clusters representing four research groups on the reliability of energy infrastructure assets, see Fig. 1.1: (i) Reliability in system design; (ii) Reliability in performance analysis; (iii) Reliability in system management; and (iv) RAM: Reliability, Availability, and Maintainability. To be specific:

(i) **Reliability in system configuration**: The research concentrates on reliability-based and reliability-related design issues. The objects are not limited to the power plant and other equipment directly involved in energy production; grid and storage facilities also occupy a considerable proportion. From the academic point of view, optimization methods, resilience, and uncertainties involved in reliability analysis, reliability estimation, reliability tests, reliability-based design optimization, and reliability improvement have been highlighted in selected studies. It is pointed out that optimization algorithms are crucial and have been widely investigated by researchers to improve the efficiency of energy production and transformation and fulfill the demand of end-users. Reliability in system configuration refers to the ability of a system to function consistently and predictably under varying conditions and workloads. It involves ensuring that

Table 1.2 Keywords analysis of the reliability of energy assets

Cluster		Keywords	Occurrences	Average citations
Code	Name			
1	Reliability in system configuration design	Optimization	42	15
		Smart grid	34	18
		System	29	21
		Energy storage	22	13
		Power system reliability	22	8
		Microgrids	20	13
		Generation	19	21
		Demand response	18	26
		Uncertainty	15	16
		Energy management	14	12
		Resilience	12	3
		Smart grids-group	10	15
		Storage	10	12
2	Reliability in performance analysis	Systems	34	14
		Energy	31	5
		Renewable energy	24	26
		Management	21	19
		Performance	15	16
		Power	15	20
		Diagnosis	12	19
		Cost	10	23
		Internet of things	10	19
3	Reliability in system management	Asset management	56	7
		Model	27	25
		Design	25	14
		Algorithm	13	21
		Distribution-systems	10	13
		Energy efficiency	10	7
4	RAM: reliability, availability, and maintainability	Reliability	127	14
		Maintenance	45	19
		Condition monitoring	14	23
		Availability	12	11
		Simulation	12	11
		Models	10	10

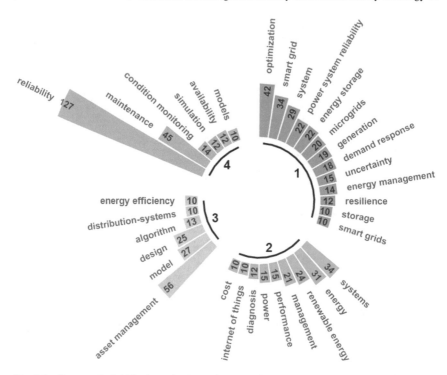

Fig. 1.1 Clusters of reliability investigations of energy infrastructure assets. 1: Reliability in system configuration; 2: Reliability in performance analysis; 3: Reliability in system management; 4: RAM

the various components and subsystems that make up a system are designed and configured to minimise the likelihood of failures or performance degradation.

To ensure reliability in system configuration, several factors should be considered during the design and configuration stages, including:

Redundancy: A system with redundant components is less likely to fail because there is always a backup component to take over in the event of a failure. Using redundant components can help ensuring that a system remains available even if a single component fails.

Scalability: A system that can scale up or down to accommodate varying workloads is more reliable than a system with a fixed capacity. Scalability allows a system to maintain performance and availability even under high load conditions.

Modularity: A system with modular components is easier to maintain and upgrade. Modularity allows for the replacement of individual parts without affecting the rest of the system.

Robustness: A robust system is designed to withstand unexpected events or environmental factors. For example, a system that is designed to handle power outages or network disruptions is more reliable than one that is not.

Testing and validation: Testing and validation are critical to ensuring that a system is reliable. Thorough testing and verifying all components and subsystems can help identify and address potential issues before they cause failures or performance degradation.

(ii) **Reliability in performance analysis**: Reliability is a crucial indicator of a system's performance. The research focuses on the functional performance and cost/economic performance of energy infrastructure assets. Failure and performance diagnosis of energy infrastructure assets are of great concern to researchers. Moreover, the diagnosis has been deepened to the system level to extract more accurate and close-to-physics features from operating energy infrastructure assets. Renewable energies like wind, solar, nuclear, ocean, etc. are interesting objects and have attracted attention from practitioners. Reliability in performance analysis refers to the ability of performance analysis tools and techniques to consistently and accurately measure the performance of a system or application. Performance analysis is the process of evaluating the performance of a system or application in terms of factors such as response time, throughput, and resource utilization.

To ensure reliability in performance analysis, several factors should be considered, including:

Validity of metrics: The metrics used to measure performance should be valid and meaningful. They should accurately reflect the performance of the system or application and be relevant to the goals of the analysis.

Consistency of measurements: Performance measurements should be consistent and repeatable. The exact size taken at different times or by other people should produce consistent results.

Accuracy of measurements: Performance measurements should be accurate and precise. The tools and techniques used to collect measurements should have a low margin of error, and any sources of measurement error should be identified and minimized.

Scope of analysis: The content of the performance analysis should be appropriate for the goals of the analysis. The study should cover all relevant aspects of the system or application but not be so broad as to be unmanageable or to produce irrelevant results.

Appropriate analysis techniques: The analysis techniques used should be appropriate for the type of system or application being analyzed. For example, the analysis techniques used for a distributed system may be different from those used for a standalone application.

Consistent analysis environment: The environment in which the analysis is conducted should be consistent across all measurements. Any environmental changes that may affect performance should be documented and accounted for.

(iii) **Reliability in system management**: The research of this group concerns the asset management of energy infrastructure assets very much. Novel models, methodologies, and algorithms are proposed to reflect the operation process of energy infrastructure assets and their management process. The research aims to guarantee energy production and transformation and improve the efficiency of the components, subsystems, units (single devices), and the entire energy system. It is to be highlighted that distribution-systems assumption is crucial for the reliability investigations of system management. Reliability in system management refers to the ability of system management practices and procedures to consistently and effectively manage a system or application. System management is the process of administering and maintaining a system or application, including tasks such as monitoring, maintenance, troubleshooting, and configuration management.

To ensure reliability in system management, several factors should be considered, including:

Consistency of management procedures: Management procedures should be consistent and repeatable. This helps ensure that the system is managed predictably and reliably, regardless of who performs the management tasks.

Monitoring and alerting: A reliable system management process requires continuous monitoring and alerting to identify issues as they arise. Automated monitoring tools can help ensure that the system is constantly monitored and that any problems are quickly identified and addressed.

Standardization: Standardization of management procedures, tools, and configurations can help improve reliability by reducing the likelihood of human error and streamlining the management process.

Documentation: Detailed and up-to-date documentation of system configuration, management procedures, and troubleshooting processes can help ensure that the system is managed consistently and effectively, even during turnover or unexpected events.

Training and expertise: A reliable system management process requires trained and experienced staff knowledgeable about the system and the management procedures in place. Training programs and ongoing professional development can ensure that the team is equipped with the knowledge and skills to manage the system effectively.

(iv) **RAM: Reliability, Availability, and Maintainability**: The research focuses on energy infrastructure assets' operation and maintenance issues. It includes all issues related to RAM of energy infrastructure assets during their life span. Among the research, failure identification and prevention, reliability modeling and analysis, availability estimation, maintenance strategy planning, and maintenance resources management are essential and highlighted in the recent research. To support the RAM of energy infrastructure assets, condition-based maintenance has attracted researchers' attention, including model-based, data-based, and model-and-data-based methodologies. RAM (Reliability, Availability, and Maintainability) is a set of three critical factors used to measure the effectiveness of a system or product. RAM focuses on the ability of a system or product to perform reliably, remain available, and be easily maintained throughout its lifecycle.

Reliability refers to the ability of a system or product to perform its intended function without failure or degradation over some time. It is usually measured in terms of Mean Time Between Failures (MTBF) or Failure Rate (FR) and is influenced by factors such as component quality, design, and manufacturing processes. Availability refers to the ability of a system or product to be available for use when needed. It is usually measured in terms of Mean Time To Repair (MTTR) or Downtime. It is influenced by reliability, maintainability, and the effectiveness of maintenance and repair processes. Maintainability refers to the ease with which a system or product can be maintained, repaired, and serviced throughout its lifecycle. It is usually measured in terms of Mean Time To Repair (MTTR) and is influenced by factors such as the system's design, the availability of spare parts, and the expertise of the maintenance staff. According to Fig. 1.1, model-based methods are the ones that practitioners have widely accepted. It is added that owing to the limited data situation when carrying out the RAM investigations of energy infrastructure assets, simulation-based methods have been applied to generate the required data for the research mentioned above activities.

Figure 1.2 displays research subtopic changes from 2014 to 2020 under the topic of the reliability of energy infrastructure assets. One can see that the hot subtopic has been changed from:

- 2014: Smart gride
- 2015: Storage
- 2016: Asset management
- 2017: Maintenance and renewable energy
- 2018: Reliability and optimization
- 2019: Design, generation, and uncertainty
- 2020: Resilience

It is concluded that the sector changed the research topics every single year. Specifically, the topics have been changed from the whole system level to specific topics. For instance, the sector focused on the reliability issues of entire devices such as storage or large-scale systems such as gride in 2014 and 2015. Subsequently, the research specified topics like asset management, maintenance, and reliability optimization in 2017–2019. More detail, uncertainties, and resilience are two aspects that have been considered. The hot topics change: From general to partial and from macro to specific issues.

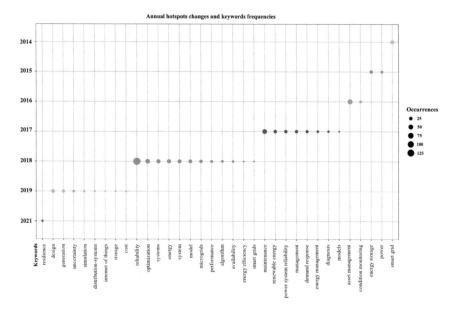

Fig. 1.2 Annual changes of hot subtopic (reliability)

It is pointed out that renewable energy has been a hot research topic since 2017. Moreover, resilience/safety/recoverability represents the concerns of the recent research.

1.3.2   Literature Review

This section reviews the detailed state-of-the-art methodologies regarding reliability investigations of energy infrastructure assets.

(1) **Devices and systems point of view**

Reliability investigations of energy infrastructure assets include risk and failure identification, as well as reliability and failure, features quantification [32]. Generally, reliability investigations are started with failure identification to ascertain the failures that may happen to energy infrastructure assets [33]. According to the failures of elements identified, the failure probability, failure rate, reliability, mean time to failure of components, systems, as well, and the entire energy infrastructure asset can be accessed based on reliability modeling tools such as, but not limited to, fault tree, reliability block diagram, Bayesian network, and Markov chain. Take the floating offshore wind turbine as an example, Li et al. [6] collected more than 400 failures from onshore wind turbines and published the latest failure data set, namely Li-Guedes Soares onshore (LGS-onshore), to determine the failures that happen to offshore wind turbines. Based on the failures collected, Li et al. [5, 6] constructed failure rate correction models, localized [6] and globalized [5], to infer the failure rates of each component of floating offshore wind turbines. Furthermore, they estimated the failure rate of an entire floating offshore wind turbine. The mentioned researches represent a research framework of reliability investigations of energy infrastructure assets and provide a practical example to any other energy infrastructure assets of other fields.

(2) **Large-scale grid systems point of view**

Regarding grid and energy system configuration and management, aiming at addressing the low flexibility, reconfigurability, and reliability of gride caused by a centralized controller, Robert et al. [34] proposed to use local information in the form of the bus voltage, which can be an alternative and effective way of improving the reliability of the entire grid. Muñoz-Delgado et al. [35] designed a multistage generation and network expansion planning for distribution systems, in which uncertainty and reliability and uncertainties of energy demand from end-users and renewable energy supplies are considered. This study proposed several alternatives for system configurations, such as feeders, transformers, and distributed generation. Beak et al. [36] constructed a secure cloud computing-based framework for smart grids, "Smart-Frame," to improve the reliability and sustainability of electricity services. The secure cloud computing-based framework is proven effective in guaranteeing grids' safety and cost-saving features. Kwon et al. [37] discussed the feasibility of implementing prognostics and systems health management technologies to assess the health state of systems, diagnose performance behavior, and predict the remaining useful life/performance over the life span of an infrastructure system based on sensors. The study concluded that analytical methodologies, sensors, and licensing approaches of the mentioned technology are the critical aspects among the others. Moghaddass and Zuo [38] proposed a model selection framework and an essential dynamic performance measure model to finalize online diagnostics and prognostics. It is concluded that the proposed prognostics and health management framework is practical and can be used to select appropriate maintenance actions to avoid catastrophic failures. It can be seen that when the reliability research step into large-scale grid system level, reliability is mainly regarded as a target or an essential factor when optimizing the configuration of energy infrastructure assets or conducting their management.

1.4 Advances in Intelligent Maintainability Investigations of Energy Infrastructure Assets

1.4.1 Bibliometric

In this part, the keywords search is completed with the topic of the maintainability of energy infrastructure assets. The minimum occurrence number of a keyword is set to 10 to exclude those keywords that are not popular and are not attractive by the recent research. Overall, 35 keywords were filtered per the composite threshold, see Table 1.3. Table 1.3 is ordered by frequency of keyword occurrence, from highest to lowest. To add more about bibliometric analysis, Bibliometric analysis is a quantitative method used to analyze and evaluate scientific literature and its impact. It involves applying statistical and computational techniques to data from various sources, such as scholarly publications, citation indexes, and databases [39, 40].

According to Table 1.3, there are four clusters representing four research groups of maintainability of energy infrastructure assets, see Fig. 1.3: (i) Condition-based maintenance; (ii) Maintenance and decision-making; (iii) RAM: Reliability, Availability and, Maintainability; and (iv) Performability. To be more specific:

(i) **Condition-based maintenance**: The research concentrates on designing, modeling, and applying condition monitoring systems and frameworks. Moreover, such a group of research is also extended to the maintenance model construction's point of view to support the maintenance strategy planning determination of energy infrastructure assets. Among other maintenance activity alternatives, predictive maintenance is investigated and applied widely. It is indicated that predicting the maintenance windows in advance of failure occurring based on the health state of energy infrastructure assets is especially important as it allows one to arrange the maintenance resources in advance and make an intelligent and cost-saving decision. On the other hand, Machine Learning and its related techniques, Digital Twin, are also becoming the main-steam techniques in the field of condition-based maintenance.

(ii) **Maintenance and decision-making**: The research concentrates on the decision-making problems in maintenance strategy planning. Preventive and condition-based maintenance are two kinds of maintenance that attract broad attention from researchers in the field. Renewable energy infrastructure assets are the primary facilities and systems investigated. Operational features such as the degradation of performance of energy infrastructure assets are the main aspects considered in the modeling process of maintenance and decision-making schedules. It is pointed out that maintenance strategy planning determination is a decision-making problem considering multiple factors like product efficiency, benefit, cost, personnel safety, etc. As a basis of that, an optimized decision can be achieved. According to the above, optimization, convincing and reliable decisions, and robust design are crucial aspects.

(iii) **RAM: Reliability, Availability, and Maintainability**: This part of the research focuses on the operation and maintenance issues of energy infrastructure assets. Like what has been done in Sect. 1.3, "Reliability in energy infrastructure assets," RAM in maintainability of energy infrastructure assets included failure identification and prevention, reliability modeling and analysis, availability estimation, maintenance strategy planning, and maintenance resources management are essential and has been highlighted in the recent research. According to Fig. 1.3, model-based methods are the ones that practitioners have widely accepted. Due to the limited data situation when carrying out the

Table 1.3 Keywords analysis of maintainability of energy assets

Cluster		Keywords	Occurrences
Code	Name		
1	Condition-based maintenance	Condition monitoring	33
		Design	33
		Systems	27
		Models	25
		Framework	23
		Predictive maintenance	21
		Diagnosis	17
		Machine learning	15
		Energy management	11
		Monitoring	11
		Digital twin	10
		Fault-detection	10
		Internet of things	10
		Networks	10
		Strategies	10
2	Maintenance and decision-making	Asset management	72
		Energy	48
		Optimization	47
		Systems	27
		Renewable energy	16
		Condition-based maintenance	15
		Degradation	14
		Generation	12
		Preventive maintenance	12
3	RAM: reliability, availability, and maintainability	Maintenance	89
		Reliability	58
		Models	17
		Operation	14
		Availability	13
		Simulation	12
4	Performability	Management	30
		Performance	26
		Sustainability	26
		Energy efficiency	18
		Cost	12

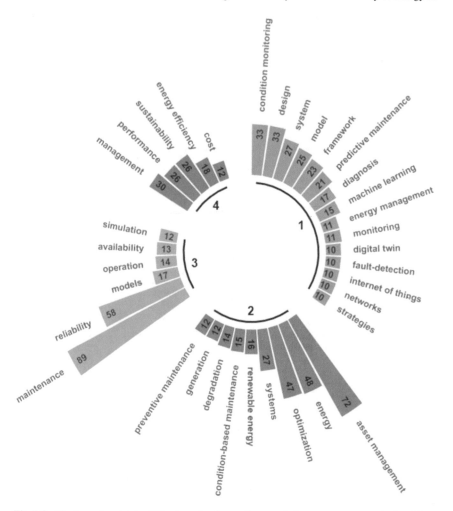

Fig. 1.3 Clusters of maintainability investigations of energy infrastructure assets. 1: Condition-based maintenance; 2: Maintenance and decision-making; 3: RAM; 4: Performability

RAM investigations of energy infrastructure assets, simulation-based methods have been applied to generate the required data for the research mentioned above activities.

(iv) **Performability**: Different from the research above-mentioned, this part of the research focuses on the effect of maintenance or, in other words, the target of implementing maintenance. For example, the researchers are finding out the best maintenance strategy to achieve the best performance of systems, including, but not limited to, sustainability of the energy generation chain, energy production efficiency, and lower cost. The research related to this part is more related to management science rather than engineering. It also indicates that maintenance/maintainability decision-making problems are multidisciplinary and complex problems.

Figure 1.4 displays research subtopic changes from 2015 to 2020 under the topic of maintainability of energy infrastructure assets. One can see that the hot subtopic has been changed from:

- 2015: Renewable energy
- 2016: Condition monitoring
- 2017: Asset management and maintenance
- 2018: Reliability and performance
- 2019: Optimization
- 2020: Machine Learning

It is concluded that the sector changed the research topics every single year. The sector focused on the renewable energy aspect in 2015. However, this research deals mainly with renewable energy generation network issues and the arrangement of maintenance activities (time and resources) for renewable energy infrastructure assets. Later on, the research deepened to condition-based maintenance to shorten down time of energy infrastructure assets so as to pull up their availability, knowing that unscheduled downtime is a factor that impacts the availability of energy infrastructure assets.

In 2017, models and tools in management science were investigated and applied to the maintenance/maintainability of energy infrastructure assets to support the maintenance strategy planning. Later on, in 2018, decision-making models of energy

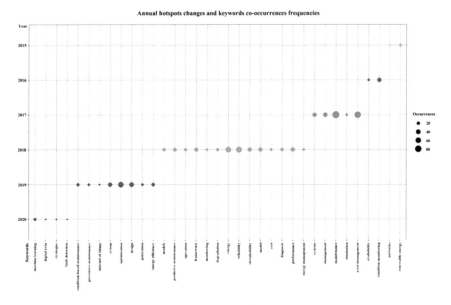

Fig. 1.4 Annual changes of hot subtopic (maintainability)

infrastructure assets considered: performance degradation, and fault diagnosis: reliability and operational state. Maintenance models were also constructed based on optimization algorithms in 2019. With the development of computer science and sensor system, machine learning was applied to the maintenance/maintainability studies of energy infrastructure assets in 2020, and this trend has lasted until today.

1.4.2 Literature Review

This section reviews the detailed state-of-the-art methodologies regarding the maintainability of energy infrastructure assets.

From the overall point of view, Shafiee and Sørensen [41] propose a conceptual classification framework for the maintenance and inspection planning for the wind energy sector. The framework addresses the models, methods, and strategies adopted to optimize inspection and maintenance decisions of the wind industry. Schneider et al. [42] extend the maintenance strategy planning to asset management during the life span of energy infrastructure assets. The research also gives an outlook on the future developments of asset management techniques. Ge et al. [43] investigated the reliability and maintainability improvement algorithms for substations with aging infrastructure. The study clarified the impacts of equipment aging, failures, and maintenance on system availability and failure frequency. It provides a valuable tool for the reliability and maintenance modeling of aging components and systems. Errandonea et al. [44] reviewed Digital Twins investigations and applications in maintenance. The study provides a framework for the new options that implement Digital Twins tools into the maintenance modeling activities. The above-selected researches indicate that the maintenance issue of energy infrastructure assets has been conducted from different perspectives.

From the specific maintenance activities point of view, preventive maintenance, corrective maintenance, and condition-based maintenance as well as their extensions have been applied to energy infrastructure assets. To be specific:

(i) **Preventive maintenance and its applications**. Daneshkhah et al. [45] proposed a probabilistic sensitivity analysis framework for preventive maintenance strategy optimization. The methodology can quantify the benefits of the maintenance strategies regarding costs and take into account the aging process of elements of energy infrastructure assets. The study contributes to the cost-effective and robust preventive maintenance planning strategy determination of the complicated systems. A critical aspect of preventive maintenance of real engineering applications, such as energy infrastructure assets, is the limitation of insufficient data. To address the limitation of the situation mentioned above, Olesen and Shaker [46] proposed a novel virtual sample generation method which is to create lifetime ratios based on the actual faults, particle swarm optimization faults, and Random Walks to generate virtual samples to enrich the data basis of preventive maintenance investigations. The proposed method provides a new way of overcoming insufficient failure information and can be used not only be applied to the maintenance of energy infrastructure assets but to their failure, risk, reliability, and availability aspects. Moreover, Wong et al. [47] reviewed the existing computational intelligence methodologies that support the intelligent preventive maintenance measures of energy infrastructure assets. The review clarified critical concerns, challenges, and opportunists using computational intelligence methodologies to support preventive maintenance strategy planning.

(ii) **Corrective maintenance and opportunistic maintenance**. Corrective maintenance is conducted after the failures or malfunctions of energy infrastructure assets have happened [48]. The main concern of this kind of maintenance is to identify failure causes and suggest related applicable maintenance activities. To this end, Wang et al. [49] proposed a complete corrective maintenance scheme based on failure modes and effect analysis and failure propagation graph. The framework can determine the most likely failure causes and appropriate maintenance actions. Recently, opportunistic maintenance emerges and allows one to conduct limited preventive maintenance after corrective maintenance. Li et al. [22] proposed a real-time opportunistic maintenance schedule based on a combination of failure modes, effect analysis, and a Bayesian network. The schedule is able to provide the failure items like systems, components, elements, and failure causes to be inspected by the maintenance crew in a real-time way and support the total operation and maintenance cost saving of the energy infrastructure assets especially floating offshore wind turbines. Opportunistic maintenance modeling also went deeper into the degradation of components and economic dependencies among multiple parts [50], reliability threshold [51], etc.

(iii) **Condition-based maintenance and predictive maintenance**. Condition-based maintenance and predictive maintenance concerns: the failures/components to be monitored, the failure diagnosis and lifetime prediction, and the decision-making approach to support the maintenance strategy determinations. Scheu et al. [52] collected 337 failure modes of offshore wind turbines. They found the most critical ones to be monitored with the assistance of the failure mode, effects, and criticality analysis method. Besnard et al. [53] presented condition-based maintenance strategies for degraded components of energy infrastructure assets. The approach applies to inspection-based maintenance based on online condition-monitoring systems. It is concluded that the proposed online condition monitoring method is able to obtain optimal maintenance strategy for components with a high failure rate which would bring additional benefit to the whole system. Predictive maintenance is a relatively new concept and tends to bring benefit to the industry as it can make smart-enough and overall optimized maintenance decisions based on all relevant information of energy infrastructure assets. Efthymiou et al. [54] propose an integrated predictive maintenance platform consisting of data acquisition and analysis, knowledge management, and a sustainability maintenance dashboard module. The platform can provide in-time guidance for maintenance activity determination. More detailed information on condition-based and predictive maintenance can be accessed in [55–59].

1.5 Discussions

The readers of this book should realize that the reliability and maintainability of energy infrastructure assets have been investigated, and several aspects of the mentioned research topics have been well addressed. However, understanding the entire framework of failure, safety, reliability, availability, and maintainability and their connections are essential for product design and the operation and maintenance of complicated systems such as energy infrastructure assets.

Undoubtedly, the methods, tools, and ideas that have been proposed and applied to similar engineering cases can be used as reference examples for energy infrastructure assets. However, the unique features of such asserts should be highlighted when applying the existing methodologies. On the other hand, the revolutionary development and application of new materials, novel designs, state of the art operation tools lead to new features of recently designed and manufactured energy infrastructure assets. Thus, the existing models may not reflect precisely the fundamental operation properties and performance of the mentioned assets.

Regarding the maintainability of energy infrastructure assets, the fowling aspects should be considered:

- **Failure collection and potential failure identification**. For operating systems, failure data collection guidelines should be constructed and based on which to collect operation data and failure data for all sectors, including failure and risk assessment sector, reliability sector, maintenance sector, manager, and other stockholders; For the systems under design, potential failures that may happen to the system/equipment should be clarified to support the robust and optimal design. For potential failure identification, failure transformation from similar products and systems is promising. Failure collection and potential failure identification are important steps in the failure analysis process, which is a systematic approach to identifying the root cause of failures and developing solutions to prevent them from recurring. Failure collection involves gathering data about the failure, including when and where it occurred, the symptoms, and any relevant contextual information. This data can be collected through various means, such as interviews, surveys, or by examining physical artefacts or data logs. Potential failure identification involves analysing the data collected to identify the underlying causes of the failure. This can include using tools such as root cause analysis, fault tree analysis, or failure mode and effects analysis. These methods help identify the sequence of events or factors that led to the failure and any contributing factors or root causes. The information gathered through failure collection and potential failure identification can then be used to develop strategies to prevent similar failures in the future. These strategies include design modifications, changes to operating procedures, or changes to maintenance practices. The failure analysis process is iterative, and the solutions developed should be evaluated over time to ensure their effectiveness in preventing future failures.

- **Failure data analysis and storage**. A standard of failure data arrangement and storage is needed as the operation and failure data accumulate fast. Failure data storage standardization supports fast and low-cost reading and application, as well as data transformation and movement among different sectors. It is pointed out that for the sectors that have already accumulated vast among of failure or operation records, natural language processing methodologies are of great need as they require excessive labor to finalize the data analysis and introduce additional costs to engineering projects. Failure data analysis and storage are important aspects of the failure analysis process. Failure data analysis involves the examination of failure data to identify trends, patterns, and potential causes of failure. This analysis can help to identify areas where improvements can be made to prevent future failures. Failure data can be collected from a variety of sources, such as maintenance logs, inspection reports, and incident reports. This data can be stored in a database or other repository for easy retrieval and analysis. The data can be categorized based on various parameters, such as the type of failure, the equipment or system involved, the location, and the time of occurrence. One approach to failure data analysis is to use statistical methods to identify trends and patterns. For example, a Pareto chart can identify the most frequent types of failures or the equipment most prone to failure. A time-series analysis can be used to identify trends in the failure rate over time. Another approach to failure data analysis is to use machine learning algorithms to identify patterns and predict future failures. This approach involves training algorithms on historical failure data and using them to predict future failures. The predictions can prioritize maintenance and inspection activities to prevent future failures. Storing failure data in a central repository can facilitate data analysis and provide a record of failures and their causes. The data can be used to inform decision-making and improve the reliability of equipment and systems. However, it is important to ensure that the data is stored securely and that sensitive information is protected.
- **Reliability issues and its improvement**. Current energy infrastructure assets tend to be intelligent, compact, and complex systems interacting with the environment and people. Hence, reliability investigations should consider and reflect the interaction/correlation among elements, the impact on working conditions, environment, and people, and conversely. It is also pointed out that the advances of new technologies challenge the conventional understanding of system reliability features. For instance, the accumulated basic failure information, like the rate and lifetime distribution of existing systems, may not apply to the new systems. Reliability issues refer to the occurrence of failures or unexpected events in equipment or systems that can lead to downtime, increased costs, and safety hazards. Improving reliability is an ongoing process that involves identifying potential failure modes, implementing strategies to prevent or mitigate failures, and monitoring equipment and systems to ensure they remain reliable over time.
- **Maintenance, decision-making, and management**. The maintenance is limited by the resources such as labor, tools, maintenance window, etc. Maintenance

strategy support-oriented decision-making is a multi-objective optimization process, including performance, cost, benefits, etc. The understanding of maintenance strategy planning should be upgraded to the resource management level from an overall point of view. On the other hand, failure, reliability, and maintainability should be considered synchronously when addressing a fundamental engineering problem. Maintenance, decision-making, and management are closely related to ensuring the reliability and availability of equipment and systems. Effective maintenance decision-making and management can help improve equipment and systems' performance and longevity and reduce costs associated with downtime and repairs.

Here are some key considerations when it comes to maintenance, decision-making, and management:

1. Maintenance strategies: Different maintenance strategies can be used, depending on the type of equipment and system and the risk associated with failure. These strategies can include reactive maintenance (fixing equipment after it fails), preventive maintenance (maintaining equipment on a regular schedule), or condition-based maintenance (maintaining equipment based on its condition or performance).
2. Risk assessment: When making maintenance decisions, it's essential to consider the risks associated with equipment failure, as well as the potential costs of maintenance and repair. A risk assessment can be used to prioritize maintenance activities and allocate resources effectively.
3. Asset management: Effective asset management involves monitoring and tracking the performance and condition of equipment and systems and planning for maintenance and replacement over the long term. This can include using software tools to track equipment performance and condition, as well as the development of maintenance plans and schedules.
4. Decision-making tools: Various decision-making tools can support maintenance decision-making, such as decision trees, cost–benefit analysis, and Monte Carlo simulations. These tools can help to evaluate the costs and benefits of different maintenance strategies and prioritize maintenance activities.
5. Continuous improvement: Effective maintenance decision-making and management involve an ongoing improvement process. Performance metrics should be tracked, and data should be analyzed to identify opportunities for improvement in maintenance practices.

1.6   Conclusions

This chapter provided a brief framework of reliability and maintainability investigations of energy infrastructure assets from a bibliometric and literature review point of view. The purpose of this book is presented. A brief introduction to the need for reliability and maintainability of energy infrastructure assets is introduced. Subsequently, details of advances and state of the art of reliability and maintainability investigations of energy infrastructure assets are described respectively. The potential opportunities for related research and engineering applications are also provided.

References

1. Omer, A.M.: Energy, environment and sustainable development. Renew. Sustain. Energy Rev. **12**(9), 2265–2300 (2008)
2. Jiang, G.J., Huang, C.G., Nedjati, A., et al. Discovering the sustainable challenges of biomass energy: a case study of Tehran metropolitan. Environ. Dev. Sustain. **8** (2023)
3. Nedjati, A., Yazdi, M., Abbassi, R.: A sustainable perspective of optimal site selection of giant air-purifiers in large metropolitan areas. Environ. Dev. Sustain. **24**, 8747–8778 (2022)
4. Cho, E.: Making reliability reliable: a systematic approach to reliability coefficients. Organ. Res. Methods **19**(4), 651–682 (2016)
5. Li, H., Soares, C.G.: Assessment of failure rates and reliability of floating offshore wind turbines. Reliab. Eng. Syst. Saf. **228**, 108777 (2022)
6. Li, H., Peng, W., Huang, C.G., Guedes Soares, C.: Failure rate assessment for onshore and floating offshore wind turbines. J. Marine Sci. Eng. **10**(12), 1965 (2022)
7. Smith, D.J.: Reliability, maintainability and risk: practical methods for engineers. Butterworth-Heinemann (2021)
8. Gardoni, P. (ed.). (2017). *Risk and Reliability Analysis: Theory and Applications*, p. 556. Springer Nature.
9. Leung, L.: Validity, reliability, and generalizability in qualitative research. J. Family Med. Prim. Care **4**(3), 324 (2015)
10. Coit, D.W., Zio, E.: The evolution of system reliability optimization. Reliab. Eng. Syst. Saf. **192**, 106259 (2019)
11. Li, H., Teixeira, A.P., Guedes Soares, C.: An improved failure mode and effect analysis of floating offshore wind turbines. J. Marine Sci. Eng. **10**(11), 1616 (2022)
12. Li, H., Yazdi, M., Huang, C.G., Peng, W.: A reliable probabilistic risk-based decision-making method: Bayesian technique for order of preference by similarity to ideal solution (B-TOPSIS). Soft. Comput. **26**(22), 12137–12153 (2022)
13. Li, H., Yazdi, M., Huang, H.-Z., Huang, C.-G., Peng, W., Nedjati, A., Adesina, K.A.: A fuzzy rough copula Bayesian network model for solving complex hospital service quality assessment. Complex Intell. Syst. (2023). https://doi.org/10.1007/s40747-023-01002-w.
14. Peyghami, S., Palensky, P., Blaabjerg, F.: An overview on the reliability of modern power electronic based power systems. IEEE Open J. Power Electron. **1**, 34–50 (2020)
15. Li, H., Yazdi, M.: An advanced TOPSIS-PFS method to improve human system reliability. In: Advanced Decision-Making Methods and Applications in System Safety and Reliability Problems, pp. 109–125. Springer, Cham (2022)
16. Li, H., Díaz, H., Soares, C.G.: A failure analysis of floating offshore wind turbines using AHP-FMEA methodology. Ocean Eng. **234**, 109261 (2021)
17. Li, H., Diaz, H., Soares, C.G.: A developed failure mode and effect analysis for floating offshore wind turbine support structures. Renew Energy **164**, 133–145 (2021)
18. Velmurugan, R.S., Dhingra, T.: Maintenance strategy selection and its impact in maintenance function: a conceptual framework. Int. J. Oper. Prod. Manage. (2015)
19. Ren, Z., Verma, A.S., Li, Y., Teuwen, J.J., Jiang, Z.: Offshore wind turbine operations and maintenance: a state-of-the-art review. Renew. Sustain. Energy Rev. **144**, 110886 (2021)
20. Li, H., Soares, C.G., Huang, H.Z.: Reliability analysis of a floating offshore wind turbine using Bayesian Networks. Ocean Eng. **217**, 107827 (2020)
21. Ben-Daya, M., Kumar, U., Murthy, D.P.: Introduction to maintenance engineering: modelling, optimization and management. Wiley (2016)

22. Li, H., Huang, C.G., Soares, C.G.: A real-time inspection and opportunistic maintenance strategies for floating offshore wind turbines. Ocean Eng. **256**, 111433 (2022)
23. Bokrantz, J., Skoogh, A., Berlin, C., Wuest, T., Stahre, J.: Smart Maintenance: a research agenda for industrial maintenance management. Int. J. Prod. Econ. **224**, 107547 (2020)
24. Zhang, Y., Andrews, J., Reed, S., Karlberg, M.: Maintenance processes modelling and optimisation. Reliab. Eng. Syst. Saf. **168**, 150–160 (2017)
25. Podlesnik, C.A., Kelley, M.E., Jimenez-Gomez, C., Bouton, M.E.: Renewed behavior produced by context change and its implications for treatment maintenance: a review. J. Appl. Behav. Anal. **50**(3), 675–697 (2017)
26. Varkevisser, R.D.M., Van Stralen, M.M., Kroeze, W., Ket, J.C.F., Steenhuis, I.H.M.: Determinants of weight loss maintenance: a systematic review. Obes. Rev. **20**(2), 171–211 (2019)
27. Shin, J.H., Jun, H.B.: On condition based maintenance policy. J. Comput. Des. Eng. **2**(2), 119–127 (2015)
28. Carvalho, T.P., Soares, F.A., Vita, R., Francisco, R.D.P., Basto, J.P., Alcalá, S.G.: A systematic literature review of machine learning methods applied to predictive maintenance. Comput. Ind. Eng. **137**, 106024 (2019)
29. Cavalcante, C.A., Lopes, R.S.: Multi-criteria model to support the definition of opportunistic maintenance policy: a study in a cogeneration system. Energy **80**, 32–40 (2015)
30. Erguido, A., Márquez, A.C., Castellano, E., Fernández, J.G.: A dynamic opportunistic maintenance model to maximize energy-based availability while reducing the life cycle cost of wind farms. Renew. Energy **114**, 843–856 (2017)
31. Li, H., Yazdi, M.: Developing failure modes and effect analysis on offshore wind turbines using two-stage optimization probabilistic linguistic preference relations. In: Advanced Decision-Making Methods and Applications in System Safety and Reliability Problems. Studies in Systems, Decision and Control, vol. 211. Springer, Cham (2022). https://doi.org/10.1007/978-3-031-07430-1_4
32. Li, H., Soares, C.G.: Reliability analysis of floating offshore wind turbines support structure using hierarchical Bayesian network. In: Proceedings of the 29th European Safety and Reliability Conference, pp. 2489–2495. Research Publishing Services Singapore (2019)
33. Li, H., Yazdi, M.: Reliability analysis of correlated failure modes by transforming fault tree model to Bayesian network: a case study of the MDS of a CNC machine tool. In: Advanced Decision-Making Methods and Applications in System Safety and Reliability Problems. Studies in Systems, Decision and Control, vol. 211. Springer, Cham (2022). https://doi.org/10.1007/978-3-031-07430-1_2
34. Balog, R.S., Weaver, W.W., Krein, P.T.: The load as an energy asset in a distributed DC smartgrid architecture. IEEE Trans. Smart Grid **3**(1), 253–260 (2011)
35. Muñoz-Delgado, G., Contreras, J., Arroyo, J.M.: Multistage generation and network expansion planning in distribution systems considering uncertainty and reliability. IEEE Trans. Power Syst. **31**(5), 3715–3728 (2015)
36. Baek, J., Vu, Q.H., Liu, J.K., Huang, X., Xiang, Y.: A secure cloud computing based framework for big data information management of smart grid. IEEE Trans. Cloud Comput. **3**(2), 233–244 (2014)
37. Kwon, D., Hodkiewicz, M.R., Fan, J., Shibutani, T., Pecht, M.G.: IoT-based prognostics and systems health management for industrial applications. IEEE Access **4**, 3659–3670 (2016)
38. Moghaddass, R., Zuo, M.J.: An integrated framework for online diagnostic and prognostic health monitoring using a multistate deterioration process. Reliab. Eng. Syst. Saf. **124**, 92–104 (2014)
39. Yazdi, M., Mohammadpour, J., Li, H., Huang, H.-Z., Zarei, E., Pirbalouti, R.G., Adumene, S.: Fault tree analysis improvements: a bibliometric analysis and literature review. Qual. Reliab. Eng. Int. (2023). https://doi.org/10.1002/qre.3271
40. Yazdi, M.: A review paper to examine the validity of Bayesian network to build rational consensus in subjective probabilistic failure analysis. Int. J. Syst. Assur. Eng. Manag. **10**, 1–18 (2019). https://doi.org/10.1007/s13198-018-00757-7

41. Shafiee, M., Sørensen, J.D.: Maintenance optimization and inspection planning of wind energy assets: Models, methods and strategies. Reliab. Eng. Syst. Saf. **192**, 105993 (2019)
42. Schneider, J., Gaul, A.J., Neumann, C., Hogräfer, J., Wellßow, W., Schwan, M., Schnettler, A.: Asset management techniques. Int. J. Electr. Power Energy Syst. **28**(9), 643–654 (2006)
43. Ge, H., Asgarpoor, S.: Reliability and maintainability improvement of substations with aging infrastructure. IEEE Trans. Power Delivery **27**(4), 1868–1876 (2012)
44. Errandonea, I., Beltrán, S., Arrizabalaga, S.: Digital Twin for maintenance: a literature review. Comput. Ind. **123**, 103316 (2020)
45. Daneshkhah, A., Stocks, N.G., Jeffrey, P.: Probabilistic sensitivity analysis of optimised preventive maintenance strategies for deteriorating infrastructure assets. Reliab. Eng. Syst. Saf. **163**, 33–45 (2017)
46. Olesen, J.F., Shaker, H.R.: Predictive maintenance within combined heat and power plants based on a novel virtual sample generation method. Energy Convers. Manage. **227**, 113621 (2021)
47. Wong, S.Y., Ye, X., Guo, F., Goh, H.H.: Computational intelligence for preventive maintenance of power transformers. Appl. Soft Comput. **114**, 108129 (2022)
48. Rausand, M., Vatn, J.: Reliability centred maintenance. In: Complex System Maintenance Handbook, pp. 79–108. Springer, London (2008)
49. Wang, Y., Deng, C., Wu, J., Wang, Y., Xiong, Y.: A corrective maintenance scheme for engineering equipment. Eng. Fail. Anal. **36**, 269–283 (2014)
50. Li, M., Wang, M., Kang, J., Sun, L., Jin, P.: An opportunistic maintenance strategy for offshore wind turbine system considering optimal maintenance intervals of subsystems. Ocean Eng. **216**, 108067 (2020)
51. Zhang, C., Gao, W., Guo, S., Li, Y., Yang, T.: Opportunistic maintenance for wind turbines considering imperfect, reliability-based maintenance. Renew. Energy **103**, 606–612 (2017)
52. Scheu, M.N., Tremps, L., Smolka, U., Kolios, A., Brennan, F.: A systematic failure mode effects and criticality analysis for offshore wind turbine systems towards integrated condition based maintenance strategies. Ocean Eng. **176**, 118–133 (2019)
53. Besnard, F., Bertling, L.: An approach for condition-based maintenance optimization applied to wind turbine blades. IEEE Trans. Sustain. Energy **1**(2), 77–83 (2010)
54. Efthymiou, K., Papakostas, N., Mourtzis, D., Chryssolouris, G.: On a predictive maintenance platform for production systems. Procedia CIRP **3**, 221–226 (2012)
55. Yazdi, M., Khan, F., Abbassi, R., Rusli, R.: Improved DEMATEL methodology for effective safety management decision-making. Saf. Sci. **127**, 104705 (2020). https://doi.org/10.1016/j.ssci.2020.104705
56. Li, X., Han, Z., Yazdi, M., Chen, G.: A CRITIC-VIKOR based robust approach to support risk management of subsea pipelines, Appl. Ocean Res. **124** (2022) 103187. https://doi.org/10.1016/j.apor.2022.103187
57. Yazdi, M.: Linguistic Methods Under Fuzzy Information in System Safety and Reliability Analysis. Springer, Cham (2022). https://doi.org/10.1007/978-3-030-93352-4%0A%0A.
58. Li, H., Guo, J.-Y., Yazdi, M., Nedjati, A., Adesina, K.A.: Supportive emergency decision-making model towards sustainable development with fuzzy expert system. Neural Comput. Appl. **33**, 15619–15637 (2021). https://doi.org/10.1007/s00521-021-06183-4
59. Li, H., Yazdi, M.: A holistic question: is it correct that decision-makers neglect the probability in the risk assessment method? In: Li, H., Yazdi, M. (eds.) BT—Advanced Decision-Making Methods and Applications in System Safety and Reliability Problems: Approaches, Case Studies, Multi-Criteria Decision-Making, pp. 185–189. Springer International Publishing, Cham (2022)

Chapter 2
Cutting Edge Research Topics on System Safety, Reliability, Maintainability, and Resilience of Energy-Critical Infrastructures

Abstract Over the last few years, the criticality of energy infrastructures has faced numerous global challenges. These include natural, human, process-based, and undesired events. There are many attempts conducted by scholars, practitioners, research centers, and academic institutes to introduce various frameworks and approaches to assess and improve the system reliability, maintainability, and resilience of energy infrastructure. The main goal of this chapter is to provide an overview of system safety, reliability, maintainability, and resilience frameworks of energy infrastructures that have been published in the past decades. In addition, it is essential to determine and analyze the common aspects of energy-critical infrastructures, the present challenges, and possible opportunities for future research directions. In this sense, an in-depth investigation is conducted using statistical metadata analysis. This subject discusses frontier directions and development trends to reveal the research status. Finally, a bibliometric study is undertaken to ascertain the most productive and influential researchers, research centers, and hotspot fields.

Keywords System safety · Critical infrastructure · Bibliometric analysis · Decision-making

2.1 Introduction

Energy-critical infrastructures are vital to human functionalities in everyday life and the social services provisions [1–3]. This includes various critical elements, either physically or virtually: electric and nuclear power plants, oil and gas industries, telecommunication, wind, solar, biomass, and hydrogen energy recourses, transportation (e.g., rails, bridges, pipelines, roads, tunnels, ports), and more. Thus, energy-critical infrastructures are essential for society and the community. Any damages to energy-critical infrastructures lead to catastrophic consequences to the society's health, safety, economy, and security [4–6]. That is while energy-critical infrastructures are unavoidably exposed to different types of man-made or undesired natural events such as floods, earthquakes, wildfires, explosions, hurricanes, and storms [7, 8]. As an example, several industrial accidents that occurred over the recent years

© The Author(s), under exclusive license to Springer Nature Switzerland AG 2023 25
H. Li et al., *Intelligent Reliability and Maintainability of Energy Infrastructure Assets*,
Studies in Systems, Decision and Control 473,
https://doi.org/10.1007/978-3-031-29962-9_2

worldwide [9–13], and some more natural disasters like floods [14, 15], the Tohoku earthquake and resulting tsunamis in Japan [16] and Katrina hurricanes in the US [17, 18].

In this regard, the recent undesired events have demonstrated that not all the potential hazards could accurately be predicted and adequately prevented [19, 20]. In order to manage the undesired and hazardous events impacting energy-critical infrastructures, scholars, practitioners, research centers, and academic institutes have started to direct their attention to system safety, reliability, maintainability, and resilience frameworks of the energy infrastructures [21–24]. The reasons why the three keywords (system safety, reliability, maintainability, and resilience) here are crucial in energy-critical infrastructure concepts are briefly defined as the following:

- System safety means that "*the system has met the minimum safety requirements. The efforts associated with system safety attempt to exceed these minimum compliance standards and provide the highest level of safety (i.e., the lowest level of acceptable risk) achievable for a given system*" [25].
- Reliability: meaning "*the probability that a product, system, or service will perform its intended function adequately for a specified period, or will operate in a defined environment without failure*" [26].
- Maintainability: meaning "*the probability that a failed component or system will be restored or repaired to a specified condition within a specified period or time when maintenance is performed in accordance with prescribed procedures*" [27], and
- Resilience: is "*a measure of the persistence of systems and of their ability to absorb change and disturbance and still maintain the same relationships between populations or state variables*" [28].

Moreover, in line with the growing concern for empowering and managing energy-critical infrastructure, system safety, reliability, maintainability, and resilience measurement tools have become critical issues for assessing, evaluating, and furthering the decision-making process. Therefore, the fundamental methods and operational techniques to measure and improve the system safety, reliability, maintainability, and resilience framework have attracted considerable research contributions over the last few years, including but not limited to [29–34]. A few literature reviews have been conducted on system safety, reliability, maintainability, and resilience of critical-energy infrastructure from different perspectives over the past few years. For example, Yazdi et al. [35] reviewed how to deal with the uncertainty of FTA-based system safety and reliability methods. Aslansefat et al. [36] also reviewed fault tree analysis methods as a powerful system and reliability analysis method modeling, analysis, and tools by distinguishing between static and dynamic fault trees. In another study [37], the authors investigated the concepts of resilience and sustainable development in group-based decision-making and conducted a systematic review. The application of the Bayesian Network as a robust tool in system safety and reliability has been comprehensively reviewed [38]. Guo et al. [39] performed a systematic literature review by identifying the main dimensions and indicators of the resilience assessment approaches utilized in critical infrastructures. However, most of the conducted review published papers can cause unreliable results and imprecise interception because they are merely based on selected papers [40, 41]. In such cases, the bibliometric analysis can add value and provide a reliable structure of advanced trends for the specified research areas. Besides, bibliometric analysis has directions using a straightforward, comprehensible, replicable, logical, and intelligible procedure [42]. Again, a unique structural approach accepts vigorous analytical methods to review published works quantitatively and reduce the bias of scholars in the available restricted works [43]. In the last few years, bibliometrics analysis has been widely used in different application domains, approaches, and fields to provide a keen understanding of the significance, properties, development path, and emerging trends of the research, including but not limited to [44–46].

This book chapter is provided to assist interested researchers, engineers, and practitioners with comprehensive system safety, reliability, maintainability, and resilience frameworks of the energy infrastructures improvement research outlook. In addition, it provides facilities to seek worldwide cooperation opportunities with other research centers and scholars and highlights the new direction for future research activities. The main contributions of the present chapter to the system reliability, maintainability, and resilience frameworks of energy-critical infrastructures literature are outlined as the following questions:

- Question 1: How did system safety, reliability, maintainability, and resilience of energy-critical infrastructures-based publication distribution change over time?
- Question 2: Which journals did contribute the most to system safety, reliability, maintainability, and resilience of energy-critical infrastructures-based research?
- Question 3: Which institutes and scholars were the most influential and prolific in system safety, reliability, maintainability, and resilience of energy-critical infrastructures-based improvement frameworks?
- Question 4: What were the most highly cited references in system safety, reliability, maintainability, and resilience of energy-critical infrastructures-based improvement frameworks?
- Question 5: What were the development path and emerging trends of system safety, reliability, maintainability, and resilience of energy-critical infrastructures-based improvement frameworks?
- Question 6: What was the contribution of different financial sectors in the system safety, reliability, maintainability, and resilience of energy-critical infrastructures-based improvement framework areas?

The present chapter is structured in the following order. Section 2.2 proposes the research methodology to conduct the literature review. Section 2.3 presents the outcomes of the bibliometric analysis performed on the collected dataset. Finally, in Sect. 2.4, conclusions and directions for future studies are discussed.

2.2 Research Methodology

The critical issue that we are following in the current overview work is performing an assessment and existing literature in energy-critical infrastructures. In this regard, the key objective of the present work is to examine the shortcoming of the state of the arts and provide reliable and valuable insights for scholars, practitioners, research centers, and academic institutes. There are a limited number of solid review work approaches. As an example, a five-stage framework considered question preparations, investigation locations, evaluating and selecting the study, analyzing and synthesizing, and presenting the outcome reports [47, 48]. In the present review work, we utilized a four stages framework to conduct the bibliometric analysis of the state of the arts in developing energy-critical infrastructures. The used framework consists of three main stages (i) defining the keywords, (ii) collecting the data, (iii) analyzing the metadata statistically, and using bibliometrics.

(i) Defining the keywords

It is essential to adopt a unique structure of keywords to assess the large scale of research terms. It also needed a reliable and comprehensive literature review coverage related to energy-critical infrastructures. According to this point, the two-level structure of keyword searching is presented as the following:

- Core keywords: "energy infrastructure" OR "critical energy infrastructure" OR "critical-energy infrastructure," AND
- Secondary keywords: "System Safety" OR "Reliability" OR "Maintainability" OR "Resilience".

It should be added that the "*core keywords*" indicate the search context, and the "*secondary keywords*" restrict the research scope.

(ii) Collecting the data

Both "*core keywords*" and "*secondary keywords*" are used for searching purposes in the Scopus database, considering the following constraints: (i) "Topic, Title, Keywords, or Abstract" is considered, (ii) the period is by the end of November 2022, (iii) the language is limited to English, (iv) source type is left as "Journal, Conference proceeding, Book series, and Book", (v) document type is selected based on "Article, Conference paper, Review paper, Book chapter, Book, and Conference review", and (vi) the publication stage is considered both studies that are in "Final" or "Article in press" stage. The initial screening resulted in 523 documents. The abstracts, keywords, and titles of the documents are then carefully reviewed. In the second screening round, 446 documents were left. The criteria for selections are: (i) the documents that applied FTA in different application domains for risk assessment purposes were retained; (ii) the documents that explained system reliability, maintainability, and resilience of critical-energy infrastructure with a simple example were excluded; (iii) the documents that focused on the general discussion of system reliability, maintainability, and resilience of critical-energy infrastructure were acknowledged as irrelevant; and (iv) the articles that aimed at extending system reliability, maintainability, and resilience of critical-energy infrastructure to deal with its shortcomings were included.

(iii) Analyzing the metadata statistically and using bibliometrics

The publication distribution based on the period by the end of November 2022 is illustrated in Fig. 2.1. It is crystal clear that the trend of distribution has slowly increased over the last decade. That is while a noticeable growth occurred in 2018 and 2022. In addition, the number of published papers in the previous three years is 137 (2019–2022), comprising 30.7% of the considered documents. It should be noted that the system reliability, maintainability, and resilience of critical-energy infrastructure improvement topic widely attracted scholars, engineers, and practitioners in recent years. In addition, the published documents contributed to the more than 22 subject areas in system reliability, maintainability, and resilience of critical-energy infrastructure improvement topics. As can be seen from Fig. 2.2, the subjects of Engineering (212 documents, 24.8%), energy (193 documents, 22.5%), and Environmental Sciences (93 documents, 10.9%) have received the three highest priority attention. This is followed by the subject of Computer Science (84 documents, 9.8%), Social Sciences (67 papers, 7.8%), and Mathematics (32 documents, 3.7%).

It is worth mentioning that metadata/bibliometric analysis is a solid and reliable approach to measuring and evaluating the impact of published papers by highlighting the direction of future research works. The analysis can be carried out using crucial information such as authors, citations, keywords, titles, journals, publication years, and more [49]. In this study, the VOSviewer software developed by Van Eck and Waltman [50] is used to conduct the bibliometric analysis for detecting, analyzing, and visualizing the trend and patterns of published scientific articles in the system

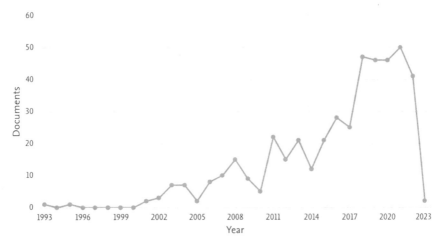

Fig. 2.1 The number of published documents over time

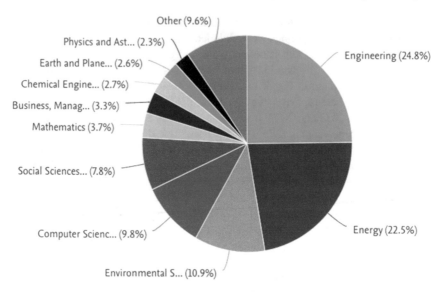

Fig. 2.2 The category of published documents based on subject areas

reliability, maintainability, and resilience of critical-energy infrastructure improvement areas. The details of the overview analysis, results, highlighting remarks, and discussions are explained in the next section.

2.3 Results and Discussions

In this section, four types of bibliometric analysis are conducted. First, the co-authorship is examined based on worldwide countries' contributions and networks. Second, the contribution of research centers and their international network was evaluated. Third, the authorship contribution based on the total documents and citations is assessed. Fourth, the overview is conducted to identify the bibliographic coupling based on reoccurs. Finally, an assessment is carried out to determine the co-occurrence of the most influential *keywords* in the system reliability, maintainability, and resilience of critical-energy infrastructure improvement areas.

- *Assessing the worldwide countries' contributions*

As seen in Fig. 2.3, the United States, United Kingdom, and China, with the published number of documents 158, 131, and 130, respectively, are the three main highly contributed countries in the system reliability, maintainability, and resilience of critical-energy infrastructure improvement areas. In addition, the total strength link of the United States, United Kingdom, and China are 53, 32, and 23 in exact accordance.

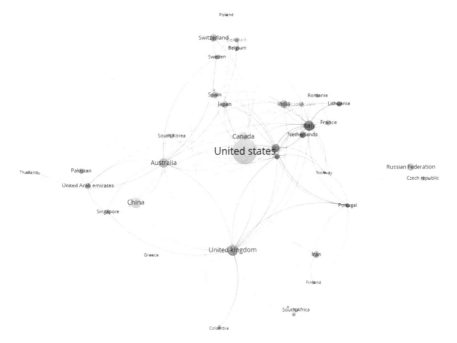

Fig. 2.3 The worldwide countries' contributions

- *Evaluating the research centers contributions and their international network.*

The results indicated that the two research centers, as Department of civil and environmental engineering, the university of western Ontario, London, Canada, and Shanghai key laboratory of financial information technology, Shanghai university of finance and economics, shanghai, China, with equal total link of 12 has the greatest world contributions to this field (Fig. 2.4).

- *The authorship's contribution is based on the total documents and citations.*

Figure 2.5 shows the highly prolific authors in this subject. The author's analysis showed that the three attractive authors in terms of published documents are Samaneh Pazouki, Mahmoudreza Haghifam, and Phillip L. Smith. Besides, Fig. 2.6 depicts the highly cited authors in this area. It can be seen that Zhang X has received 619 citations, with a total strength link of 19. It is followed by Shahidehpour, M and Abusorrah, A with equal total citations of 548 and a total strength link of 7.

- *Identifying the bibliographic coupling based on recourses*

It should be added that the "two publications are bibliographically coupled if there is a third publication that is cited by both publications" [51]. Similarly, the "bibliographic coupling is about the overlap in the reference lists of publications" [52]. Figure 2.7

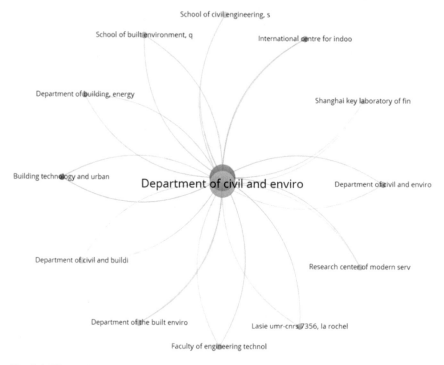

Fig. 2.4 The worldwide research centers contributions and their international network

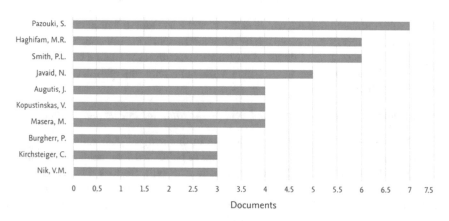

Fig. 2.5 The highly prolific authors

depicts that the three highest impactful recourses based on bibliographic coupling are Journals of Reliability Engineering and System Safety with a total strength link of 140, IEEE transaction smart grid with a total strength link of 97, and Energies with a total strength link of 87.

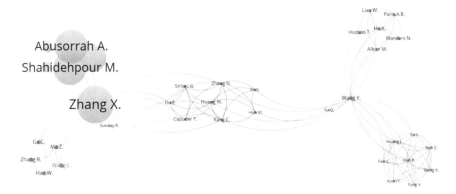

Fig. 2.6 The highly cited authors

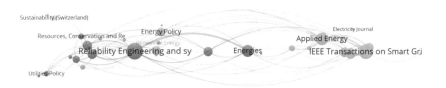

Fig. 2.7 The bibliographic coupling analysis based on recourses

- *Assessing the co-occurrence of the most influential keywords*

As can be seen from Fig. 2.8, the co-occurrence of most influential keywords based on "All Keywords" and considering the minimum number of occurrences equal to 8 is: "Energy infrastructure" with the event of 171 and total strength link of 733, "Electric power transmission network" with the occurrence of 40 and total strength link of 181, and "Reliability" with the occurrence of 36 and total strength link of 157. It is then followed by "Climate change" (35, 118), "Investments" (34, 228), and "Energy security" (34, 182). In addition, Fig. 2.9 illustrates the co-occurrence of the most influential keywords based on "Authors Keywords" and considers the minimum occurrence number equal to 2. The result indicated the most significant "Authors Keywords" are "Resilience," with the occurrence of 27 and a total strength link of 44, "Energy infrastructure," with the occurrence of 27 and a total strength link of 37, and "Renewable energy" with the occurrence of 23 and total strength link of 34. It is then followed by "Smart grid" (19, 48), "Energy security" (12, 27), and "Climate change" (11, 23).

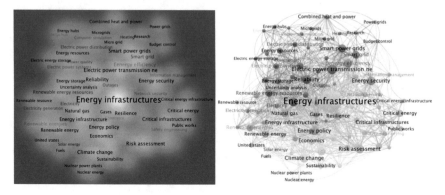

Fig. 2.8 The co-occurrence of the most influential "all keywords"

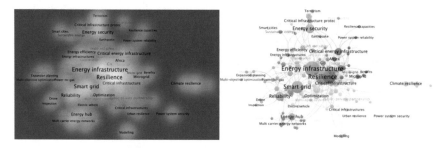

Fig. 2.9 The co-occurrence of most influential "authors keywords"

2.4 Conclusion

According to the 446 collected and identified documents in this preset review, a bibliometric analysis of the system reliability, maintainability, and resilience of critical-energy infrastructure improvement areas is conducted. The main findings and significant conclusions are summarized as the following:

- In terms of affiliation analysis, it can be concluded the United States, the United Kingdom, and China are the key contributor to this topic,
- The Department of civil and environmental engineering, the university of western Ontario, in Canada, and Shanghai key laboratory of financial information technology in China was identified as the top two contributing research centers,
- The National Natural Science Foundation of China, the National Science Foundation, and the U.S. Department of Energy were identified as the top three funding sponsors,
- The publication distributions analysis indicated that the Journals of Reliability Engineering and System Safety, IEEE transaction on smart grid, and Energies were the leading journals on system reliability, maintainability, and resilience of critical-energy infrastructure improvement topics,

- The author's analysis showed that the three attractive authors in terms of published documents are Samaneh Pazouki, Mahmoudreza Haghifam, and Phillip L. Smith,
- Besides, Zhang X is identified as a highly cited scholar in this field,
- In terms of keywords co-occurrence analysis, it was revealed that the "Energy infrastructure," "Electric power transmission network," "Reliability," "Climate change," "Investments," and "Energy security" were the hotspots in this field,
- The subject area analysis indicates that the improved system reliability, maintainability, and resilience of critical energy was commonly applied to Engineering, Energy, Environmental Sciences, Computer Science, Social Sciences, and Mathematics.

However, some challenges were faced during the study, which required appropriately handled as a direction for future studies. These are summarized as the following:

- The VOSviewer software in this study does not have much capacity to show the dynamic behavior of published documents. In future works, it is suggested to use to utilize CiteSpace or develop in-house codes, and
- In this review work, decision-makers subjectivity plays a significant role in studying all related published documents. In this regard, a comprehensive reference analysis structure is suggested to be employed in further works.

References

1. Greiving, S., Fleischhauer, M., León, C.D., Schödl, L., Wachinger, G., Quintana Miralles, I.K., et al.: Participatory assessment of multi risks in urban regions—the case of critical infrastructures in metropolitan lima. Sustainability **13**, 2813 (2021). https://doi.org/10.3390/su1305 2813
2. Roe, E., Schulman, P.R.: Toward a comparative framework for measuring resilience in critical infrastructure systems. J. Comp. Policy Anal. Res. Pract. **14**, 114–125 (2012). https://doi.org/10.1080/13876988.2012.664687
3. Yazdi, M., Kabir, S.: Fuzzy evidence theory and Bayesian networks for process systems risk analysis. Hum. Ecol. Risk Assess. **26**, 57–86 (2020). https://doi.org/10.1080/10807039.2018.1493679
4. Hassan, E.M., Mahmoud, H.: An integrated socio-technical approach for post-earthquake recovery of interdependent healthcare system. Reliab. Eng. Syst. Saf. **201**, 106953 (2020). https://doi.org/10.1016/j.ress.2020.106953
5. Yoo, B.H., Wilailak, S., Bae, S.H., Gye, H.R., Lee, C.J.: Comparative risk assessment of liquefied and gaseous hydrogen refueling stations. Int. J. Hydrogen Energy **46**, 35511–35524 (2021). https://doi.org/10.1016/j.ijhydene.2021.08.073
6. Kalantarnia, M., Khan, F., Hawboldt, K.: Modelling of BP Texas City refinery accident using dynamic risk assessment approach. Process. Saf. Environ. Prot. **88**, 191–199 (2010). https://doi.org/10.1016/j.psep.2010.01.004
7. Argyroudis, S.A., Mitoulis, S.A., Hofer, L., Zanini, M.A., Tubaldi, E., Frangopol, D.M.: Resilience assessment framework for critical infrastructure in a multi-hazard environment: case study on transport assets. Sci. Total Environ. **714**, 136854 (2020). https://doi.org/10.1016/j.scitotenv.2020.136854

8. Twumasi-Boakye, R., Sobanjo, J.O.: Resilience of regional transportation networks subjected to hazard-induced bridge damages. J. Transp. Eng. Part A Syst. **144**, 4018062 (2018). https://doi.org/10.1061/JTEPBS.0000186

9. Li, H., Yazdi, M.: Stochastic game theory approach to solve system safety and reliability decision-making problem under uncertainty. In: Li, H., Yazdi, M. (eds) Advanced Decision-Making Methods and Applications in System Safety and Reliability Problems: Approaches, Case Studies, Multi-criteria De. Springer International Publishing, Cham (2022), pp 127–151. https://doi.org/10.1007/978-3-031-07430-1_8

10. Yazdi, M., Adesina, K.A., Korhan, O., Nikfar, F.: Learning from fire accident at Bouali Sina petrochemical complex plant. J. Fail Anal. Prev. (2019). https://doi.org/10.1007/s11668-019-00769-w

11. Yazdi, M., Korhan, O., Daneshvar, S.: Application of fuzzy fault tree analysis based on modified fuzzy AHP and fuzzy TOPSIS for fire and explosion in process industry. Int. J. Occup. Saf. Ergon. 1–18 (2018). https://doi.org/10.1080/10803548.2018.1454636

12. Vasheghani, F.J.: Man-made major hazards like earthquake or explosion; case study, Turkish mine explosion (13 May 2014). Iran J. Public Health **43**, 1444–1450 (2014)

13. Guan, Y., Zhao, J., Shi, T., Zhu, P.: Fault tree analysis of fire and explosion accidents for dual fuel (diesel/natural gas) ship engine rooms. J. Mar. Sci. Appl. **15**, 331–335 (2016). https://doi.org/10.1007/s11804-016-1366-6

14. Li, H., Guo, J.-Y., Yazdi, M., Nedjati, A., Adesina, K.A.: Supportive emergency decision-making model towards sustainable development with fuzzy expert system. Neural Comput. Appl. **33**, 15619–15637 (2021). https://doi.org/10.1007/s00521-021-06183-4

15. Kimuli, J.B., Di, B., Zhang, R., Wu, S., Li, J., Yin, W.: A multisource trend analysis of floods in Asia-Pacific 1990–2018: implications for climate change in sustainable development goals. Int. J. Disaster Risk. Reduct. **59**, 102237 (2021). https://doi.org/10.1016/j.ijdrr.2021.102237

16. Krausmann, E., Cruz, A.M.: Impact of the 11 March 2011, Great East Japan earthquake and tsunami on the chemical industry. Nat. Hazards **67**, 811–828 (2013). https://doi.org/10.1007/s11069-013-0607-0

17. Henry, D., Ramirez-Marquez, J.E.: On the impacts of power outages during Hurricane Sandy—a resilience-based analysis. Syst. Eng. **19**, 59–75 (2016). https://doi.org/10.1002/sys.21338

18. Petterson, J.S., Stanley, L.D., Glazier, E., Philipp, J.: A preliminary assessment of social and economic impacts associated with Hurricane Katrina. Am. Anthropol. **108**, 643–670 (2006)

19. Ghasemian, H., Zeeshan, Q.: Failure mode and effect analysis (FMEA) of aeronautical gas turbine using the fuzzy risk priority ranking (FRPR) approach. Int. J. Soft. Comput. Eng. **7**, 81–92 (2017)

20. Jiang, G.-J., Chen, H.-X., Sun, H.-H., Yazdi, M., Nedjati, A., Adesina, K.A.: An improved multi-criteria emergency decision-making method in environmental disasters. Soft. Comput. (2021). https://doi.org/10.1007/s00500-021-05826-x

21. Yazdi, M., Khan, F., Abbassi, R., Quddus, N.: Resilience assessment of a subsea pipeline using dynamic Bayesian network. J. Pipeline Sci. Eng. **2**, 100053 (2022). https://doi.org/10.1016/j.jpse.2022.100053

22. Yazdi, M.: Acquiring and sharing tacit knowledge in failure diagnosis analysis using intuitionistic and Pythagorean assessments. J. Fail. Anal. Prev. **19**, 369–386 (2019). https://doi.org/10.1007/s11668-019-00599-w

23. Zhang, D., Du, F., Huang, H., Zhang, F., Ayyub, B.M., Beer, M.: Resiliency assessment of urban rail transit networks: Shanghai metro as an example. Saf. Sci. **106**, 230–43 (2018). https://doi.org/10.1016/j.ssci.2018.03.023

24. Zhu, W., Castanier, B., Bettayeb, B.: A dynamic programming-based maintenance model of offshore wind turbine considering logistic delay and weather condition. Reliab. Eng. Syst. Saf. **190**, 106512 (2019). https://doi.org/10.1016/j.ress.2019.106512

25. Vincoli, J.W.: Basic Guide to System Safety: Vincoli/Basic. Wiley, Hoboken (2014)

26. Rausand, M., Hoyland, A.: System reliability theory: models, statistical methods, and applications **664** (2004). https://doi.org/10.1109/WESCON.1996.554026

27. Gaonkar, R.S.P., Verlekar, M.V.: Reliability and maintainability of safety instrumented system. In: Pham, H., Ram, M.B.T.-S., et al. (eds). Safety and Reliability Modeling and its Applications, pp. 43–90. Elsevier, Amsterdam (2021). https://doi.org/10.1016/B978-0-12-823323-8.00005-2

28. Holling, C.S.: Resilience and stability of ecological systems. Annu. Rev. Ecol. Syst. **4**, 1–23 (1973). https://doi.org/10.1146/annurev.es.04.110173.000245

29. Li, H., Yazdi, M., Huang, H.-Z., Huang, C.-G., Peng, W., Nedjati, A., Adesina, K.A.: A fuzzy rough copula Bayesian network model for solving complex hospital service quality assessment. Complex Intell. Syst. (2023). https://doi.org/10.1007/s40747-023-01002-w.

30. Yazdi, M., Golilarz, N.A., Adesina, K.A., Nedjati, A.: Probabilistic risk analysis of process systems considering epistemic and aleatory uncertainties: a comparison study. Int. J. Uncertainty Fuzziness Knowl. Based Syst. **29**, 181–207 (2021). https://doi.org/10.1142/S0218488521500098

31. Adumene, S., Okwu, M., Yazdi, M., Afenyo, M., Islam, R., Orji, C.U., et al.: Dynamic logistics disruption risk model for offshore supply vessel operations in Arctic waters. Marit. Transp. Res. **2**, 100039 (2021). https://doi.org/10.1016/j.martra.2021.100039

32. Pouyakian, M., Khatabakhsh, A., Yazdi, M., Zarei, E.: Optimizing the allocation of risk control measures using fuzzy MCDM approach: review and application. In: Yazdi, M. (ed.) Linguistic Methods Under Fuzzy Information in System Safety and Reliability Analysis, pp. 53–89. Springer International Publishing, Cham (2022). https://doi.org/10.1007/978-3-030-93352-4_4

33. Mohammadfam, I., Zarei, E., Yazdi, M., Gholamizadeh, K.: Quantitative risk analysis on rail transportation of hazardous materials. Math. Probl. Eng. **2022**, 6162829 (2022). https://doi.org/10.1155/2022/6162829

34. Omidvar, M., Zarei, E., Ramavandi, B., Yazdi, M.: Fuzzy Bow-Tie Analysis: Concepts, Review, and Application. In: Yazdi, M. (ed.) Linguistic Methods Under Fuzzy Information in System Safety and Reliability Analysis, pp. 13–51. Springer International Publishing, Cham (2022). https://doi.org/10.1007/978-3-030-93352-4_3.

35. Yazdi, M., Kabir, S., Walker, M.: Uncertainty handling in fault tree based risk assessment: state of the art and future perspectives. Process. Saf. Environ. Prot. **131**, 89–104 (2019). https://doi.org/10.1016/j.psep.2019.09.003

36. Aslansefat, K., Kabir, S., Gheraibia, Y., Papadopoulos, Y.: Dynamic fault tree. Analysis (2020). https://doi.org/10.1201/9780429268922-4

37. Aghazadeh Ardebili, A., Padoano, E.: A literature review of the concepts of resilience and sustainability in group decision-making. Sustainability **12**, 2602 (2020). https://doi.org/10.3390/su12072602

38. Yazdi, M.: A review paper to examine the validity of Bayesian network to build rational consensus in subjective probabilistic failure analysis. Int. J. Syst. Assur. Eng. Manag. **10**, 1–18 (2019). https://doi.org/10.1007/s13198-018-00757-7

39. Guo, D., Shan, M., Owusu, E.K.: Resilience assessment frameworks of critical infrastructures: state-of-the-art review. Buildings **11**, 1–18 (2021). https://doi.org/10.3390/buildings11100464

40. Ferreira, F.A.F., Santos, S.P.: Two decades on the MACBETH approach: a bibliometric analysis. Ann. Oper. Res. **296**, 901–925 (2021). https://doi.org/10.1007/s10479-018-3083-9

41. Kazemi, N., Modak, N.M., Govindan, K.: A review of reverse logistics and closed loop supply chain management studies published in IJPR: a bibliometric and content analysis. Int. J. Prod. Res. **57**, 4937–4960 (2019). https://doi.org/10.1080/00207543.2018.1471244

42. Huang, J., You, J.X., Liu, H.C., Song, M.S.: Failure mode and effect analysis improvement: a systematic literature review and future research agenda. Reliab. Eng. Syst. Saf. **199**, 106885 (2020). https://doi.org/10.1016/j.ress.2020.106885

43. Huang, J., Mao, L.X., Liu, H.C., Song, M.: Quality function deployment improvement: a bibliometric analysis and literature review. Qual. Quant. **56**, 1347–1366 (2022). https://doi.org/10.1007/s11135-021-01179-7

44. Yazdi, M., Mohammadpour, J., Li, H., Huang, H.-Z., Zarei, E., Pirbalouti, R.G., Adumene, S.: Fault tree analysis improvements: a bibliometric analysis and literature review. Qual. Reliab. Eng. Int. (2023). https://doi.org/10.1002/qre.3271

45. Wang, X., Xu, Z., Su, S.-F., Zhou, W.: A comprehensive bibliometric analysis of uncertain group decision making from 1980 to 2019. Inf. Sci. **547**, 328–353 (2021). https://doi.org/10.1016/j.ins.2020.08.036
46. Tandon, A., Kaur, P., Mäntymäki, M., Dhir, A.: Blockchain applications in management: a bibliometric analysis and literature review. Technol. Forecast Soc. Change **166**, 120649 (2021). https://doi.org/10.1016/j.techfore.2021.120649
47. Garza-Reyes, J.A.: Lean and green—a systematic review of the state of the art literature. J. Clean Prod. **102**, 18–29 (2015). https://doi.org/10.1016/j.jclepro.2015.04.064
48. Yazdi, M., Khan, F., Abbassi, R., Quddus, N., Castaneda-Lopez, H.: A review of risk-based decision-making models for microbiologically influenced corrosion (MIC) in offshore pipelines. Reliab. Eng. Syst. Saf. **223**, 108474 (2022). https://doi.org/10.1016/j.ress.2022.108474
49. Ganbat, T., Chong, H.-Y., Liao, P.-C., Wu, Y.-D.: A bibliometric review on risk management and building information modeling for international construction. Adv. Civ. Eng. **2018**, 8351679 (2018). https://doi.org/10.1155/2018/8351679
50. van Eck, N.J., Waltman, L.: Software survey: VOSviewer, a computer program for bibliometric mapping. Scientometrics **84**, 523–538 (2010). https://doi.org/10.1007/s11192-009-0146-3
51. Kessler, M.M.: Bibliographic coupling between scientific papers. Am. Doc. **14**, 10–25 (1963). https://doi.org/10.1002/asi.5090140103
52. van Eck, N.J., Waltman, L.: Visualizing bibliometric networks. In: Ding, Y., Rousseau, R., Wolfram, D. (eds). Measuring Scholarly Impact Measuring Scholarly Impact, pp. 285–320. Springer International Publishing, Cham (2014). https://doi.org/10.1007/978-3-319-10377-8_13

Chapter 3
Operations Management of Critical Energy Infrastructure: A Sustainable Approach

Abstract The present work aims to study the application of a dynamic Bayesian modeling network to manage a critical offshore infrastructure under the influence of the material degradation process. The developed model is applied to burst probabilistic pressure modeling for safety improvement over time. The design methods suggested by ASME B31G and other empirical formulation methods are utilized to evaluate the burst pressure. The relevant parameters in the developed model are then involved in the dynamic Bayesian network modeling to determine an accurate estimation of the critical offshore infrastructure degradation and further sustainable operation management. The sensitivity analysis is carried out to investigate how changes in the input parameters can impact the outcomes and verify the robustness of the model. The results indicated that the proposed model has a high capacity to predicate the rate of material degradation, particularly under uncertain environments.

Keywords Project and operation management · Decision-making · Offshore structures · Causal modeling · Uncertainty handling

3.1 Introduction

In recent years, energy infrastructures have been the most crucial part of energy exploration, transition, and storage. However, the operational and maintenance cost of such critical infrastructures have increased worldwide and have led to a higher number of undesired events caused by material degradations (e.g., different kinds of corrosion), either internally or externally [1–3]. Theoretically, several limit state functions are considered in designing energy infrastructures and utilized for reliability analysis, including bursting and collapse caused by internal and external pressure, respectively [4–7], and buckling [8]. Several probabilistic models in the existing literature examine energy infrastructure's system safety and reliability. In this chapter, the main idea is to assess the failure mode of an energy infrastructure due to internal burst pressure.

As a common practice, regular maintenance actions such as inspections can reduce the probability of undesired events in energy infrastructures. Considering the cost of regular maintenance actions, it is not a feasible and efficient way. Suppose the system has enough capacity to update the results by receiving new evidence from inspection. In that case, it is also possible to formulate the contributions of maintenance practice to system safety and reliability of energy infrastructures. Thus, it is required to develop a time-dependent probabilistic model that could be utilized over the operation life of the energy infrastructures. There are well-established conventional reliability analysis tools (e.g., first and second-order reliability methods) in the existing state of the arts [9, 10]. They may need more potential and capacity to deal with inherent features such as time-dependent decision-making problems. Over the last decade, the Bayesian network and its extensions as a dynamic Bayesian Network, a continuous Bayesian Network, and integration with the developed fuzzy set theory have attracted the interest of scholars and practitioners in different application domains to model causality under uncertainty [11–14]. In addition, Bayesian Network enables decision-makers to model cause-and-effect problems with high complexity, many multi-states and continuous random variables, and objective/subjective uncertainty in the variable's estimations [15, 16].

Bayesian Network has successfully applied the material degradation of energy infrastructure, including Microbiologically corrosion [1–3, 17, 18], CO_2 corrosion [19], and fatigue crack growth [20]. Although the probabilistic risk assessment of energy infrastructures failures like burst transition pipelines has been broadly investigated in literature; however, timely-dependent probabilistic risk assessment has yet to receive that much attention. Thus, this calls to conduct further attempts.

In this chapter, a dynamic Bayesian Network is developed to study an energy infrastructure's time-dependent reliability, suspected to be a highly material deterioration process. It is vital to have an accurate rate of material degradation to manage energy infrastructure over time reliably. In literature, there are a couple of models for this purpose, such as but not limited to [21–23]; however, the outcomes of all studies with the same input data are different. In this accordance, it is necessary to utilize a predicted-based model with enough capacity to update itself and results once new evidence becomes available.

The organization of the present chapter is provided as the following. In Sect. 3.2, a brief explanation of the dynamic Bayesian Network and a relevant description of the material degradation process are explained. In Sect. 3.3, a demonstration of the case study is presented. Finally, a conclusion highlighting the remarks of the present study is provided.

3.2 Methodology

3.2.1 Constructing a Dynamic Bayesian Network

To build up a dynamic Bayesian Network, three and only three pieces of information are needed:

- A prior distribution for the states' variables describing the variables initially without considering any kinds of observations.
- The transition model by explaining how the states' variable evolves over the period from t = 0 to t = t−1, and
- A model for the investigations, which describes the measurement accuracy.

Once all of the information mentioned earlier is derived, the structure of the Bayesian Network can be constructed utilizing just two different time slices, one at time zero, meaning the initial level or before any observation. The second is when there is the first set of observations or when new evidence becomes available. Bayesian Network construction needs to have a proper definition of a prior distribution.

After constructing the time zero Bayesian Network, it would be enough to replicate the structure to derive the states' variables in the next time slice. As a rule of thumb, it is commonly done by adding a second structure on the right-hand side of zero time. In the next step, the links between the time slices are established. This can be conducted with consideration of the variables' history and examine how the past affects the present problem state. In the end, it is vital to include the pieces of evidence of the nodes in the constructed Bayesian Network regarding the second time slice. Generally, the evidence is considered as the child of the states' variable.

3.2.2 Dynamic Bayesian Network Inference

Using a filtering algorithm, the joint probability can be determined at a single and particular time slice, considering the last time slice. This situation is general for any probabilistic temporal model, and it is vital to conduct probabilistic inference within the dynamic Bayesian network time slice. There are different ways that decision-makers can carry out the inference in a Bayesian network. One of the most straightforward methods is "enumeration," which includes the summation of joint distributions. The distribution of a set of variables X, within a model having n state variables, Y, and several evidence variables, E, is defined as the following equation:

$$P(X|e) = \alpha P(X, e) = \alpha \sum_Y P(X, e, y) \tag{3.1}$$

In addition, several different algorithms help decision-makers avoid the operation reputations in the equation above summations by time. The utilized dynamic Bayesian network in this chapter is to use a "variable elimination: technique, which enables to store of the partial summation results and reuse them again". Inference in a dynamic Bayesian network is conducted by "unrolling" the network into a size consisting of all the evidence up to date. A graphical representation of the unrolling is provided in the literature [14].

The inference can be carried out similarly to the regular Bayesian networks procedure in the next step. In the case of filtering, the inference is initially carried out for each single time slice, and the outcomes are then stored to use and update the predictions in the future time slices based on the following equation:

$$P\left(X_{t+1}, e_{1:t+1}\right) = \alpha P(e_{t+1}|X_{t+1}) \sum_{x_t} P(X_{t+1}|x_t)P(x_t|e_{1:t}) \qquad (3.2)$$

According to this point, the filtering can be finalized with constant computational time and space for each single time slice. To get more details about Bayesian Network, dynamic Bayesian network, and temporal models, one can refer to the following references [24–26].

3.2.3 Discretizing the Variables

The main challenge of using Bayesian and dynamic Bayesian networks is dealing with hybrid and continuous models of the variables, which contain a combination of continuous and discrete random variables. Several techniques have been widely introduced to cope with such problems [27–29]. In this chapter, we assumed that all random variables are constructed as random and have been discretized to conduct the model inference. Utilizing a method for discretizing the random variable impact the accuracy of the outcomes. That is while the complexity of calculation has increased exponentially with the number of discrete variables. Thus, it is vital to find a balance between the accuracy of the outcome and the calculation complexity effort.

It should be added that in the system safety and reliability analysis problems, it is commonly studied rare events, and the failure usually occurs at the tails of random variables' probability distributions. Therefore, selecting the appropriate scheme of discretization is vital to optimizing the accuracy without increasing the partitions. In most reliability-based decision-making problems, the variables are discretized into the smaller bin in their domains, leading to the failure occurrence since the larger one should be considered when failure is less likely [30].

In this chapter, we considered the simpler approach, in which the variable's discretization area is restricted to establish a limit probability function called *pl*. The variables are then subdivided into bins with an equal length for the large area, which is enough to comprise an occurrence probability $(1 - 2pl)$. Thus, the right and left tails sides of occurrence probability distributions of *pl* left actually out. The center point for every single bin is considered a discrete value t to be utilized in performing inference and computations of the joint probability distribution over time.

3.2.4 Reliability Analysis of the Energy Infrastructure Under the Material Degradation Process

In this section, a limit state function is defined and utilized in this study to assess and analyze the reliability of energy infrastructure under material degradation for the annual internal operational pressure $P_{\max}$ and the burst pressure of the degraded infrastructure, as the following:

$$g(X) = P_b(t, D, \sigma_y, d(T, P_o, nCor), l) - P_{\max} \qquad (3.3)$$

According to the equation above, three approaches to burst pressure are applied for reliability prediction; however, all of them share the basic variables with comparable dependency connections. Therefore, in this study, a dynamic Bayesian model as an example is illustrated, and in fact, the conditional probability tables related to each node may differ from one to another.

In addition, the $g(X)$ depends on the P_b (burst pressure) and $P_{\max}$ (maximum annual internal pressure). Besides, the basic variables influence the material degradation (Cr) of energy infrastructure and P_b in the initial time (i.e., time slice $t = 0$). It should be added that the interrelationships between the nodes are deterministic. Therefore, the fixed value of parent variables in a dynamic Bayesian network demonstrates the fixed value for child ones. As mentioned earlier, in the next time slice, the basic variables depend on the values of the previse time slice. The only exception is material degradation parameters such as depth and length of corrosion.

This study considers several states' variables and a couple of evidence variables to analyze the reliability of energy infrastructure material degradation. Moreover, several models are adopted from the literature regarding probabilistic distributions [31], including yield stress, diameter, wall thickness, and internal operating pressure. Thus, the burst pressure of intact infrastructure provided by "*ASME code B31G*" in the following equations is considered 0.72.

$$P_b^{B31G} = P_{bi} \left[\frac{1 - \frac{2}{3} \cdot \frac{d}{t}}{1 - \frac{2}{3} \cdot \frac{d}{t} \cdot \frac{1}{M}} \right] \qquad (3.4)$$

$$M = \left(1 + 0.81 \cdot \left(\frac{l}{D}\right)^2 \cdot \left(\frac{D}{t}\right)\right)^{\frac{1}{2}} \qquad (3.5)$$

$$P_{bi} = \frac{1.1\sigma_y 2t}{D} \qquad (3.6)$$

According to this point, the characteristics value of operational pressure is utilized to compute the pressure impact on the material degradation rate. The extreme value distributions are then used to compute the reliability index using code DNV-RP101. It is assumed that the maximum internal pressure is based on the Gumbel distribution. The present study considers the representative values to analyze the average evolution of material degradation parameters over time. The dynamic Bayesian network deals with new updated evidence, predicts the theoretical material degradation rate as accurately as possible and enhances the reliability of degraded infrastructure.

Once the prior joint probability distributions are established, defining the transition model is vital. In this study, most of the basic variables may not be dependent and would be transferred as identified. As mentioned earlier, the only exceptions are the depth and length of material degradation, which are computed using variables in the previous time slice and the relevant degradation rate.

An investigation and inspection are required to validate the accuracy of the measurement tool. Each of them is based on the relevant state variables, considering measuring error constructed on the normal distribution with a mean value equal to zero. The rest of the parameters, such as temperature and pressure, are considered to be derived with continuous system monitoring (e.g., sensors) over a specific period. The dimensions of material degradation can be obtained with periodical inspections. The mean and standard deviation for these normal distributions can be determined from industrial practice or state of the arts.

It is worth mentioning that the utilized formulation of burst pressure and material degradation rate presented earlier are assumed to be deterministic. This is crystal clear that in case of variables are identified and have known values, the burst pressure is computed deterministically.

3.3 Application of Study

In this section, we defined an application of a case study for an energy infrastructure highly suspected of a material degradation process (i.e., corrosion). In order to comprehensively define the prior joint probability distributions, Table 3.1 presents the theoretical prior probability distributions of all contributing factors in the material degradation of understudy energy infrastructure.

Considering the input information provided in Table 3.1, the first time slice as a causality diagram is depicted in Fig. 3.1, and for the initial step, all defined variables are appropriately discretized. The widely applied tool GeNIe Modeler (https://www. bayesfusion.com) is used to precisely carry out the discretization of variables. In the next step, reliability analysis is needed when investigations and inspections are started by identifying the corrosion defect over time. This value will be known for the following years. However, it is worth mentioning that this is optional to have this information for the first year of energy infrastructure installation since a dynamic Bayesian network enables decision-makers to update the outcomes once new evidence about the corrosion parameters becomes available. In this regard, the time axis presented in the figures does not indicate the relevant service years of energy infrastructure.

In this study, we assumed that there are four and only four evidence variables, including temperature, operational pressure, and the depth and length of corrosion. The simulated value is then determined annually once the corrosion defect is detected

Table 3.1 The prior probabilistic values for the defined variables

Row	Defined variables	Distribution	Mean Approx.	Std. Approx.
#V1	Maximum annual internal pressure, P_{max}	Gumbel	19.65 MPa	0.345 MPa
#V2	Mean internal pressure P_o	Normal	16.91 MPa	0.92 MPa
#V3	Thickness, t	Normal	13.52 mm	0.162 mm
#V4	Temperature, T	Normal	42 °C	3 °C
#V5	Yield stress, σ_y	Lognormal	401.3 MPa	29.6 MPa
#V6	Diameter, D	Normal	418.2 mm	0.387 mm
#V7	CO_2 concentration, pCO_2	Lognormal	0.15 mol %	0.02 mol %
#V8	Longitudinal corrosion rate, CR	Lognormal	18 mm per year	9 mm per year
#V9	Corrosion parameter: depth, d	Deterministic	0 mm[a]	N/A
#V10	Corrosion parameter: length, l	Deterministic	0 mm[a]	N/A

[a] It is assumed to be 0 mm at the initial time slice

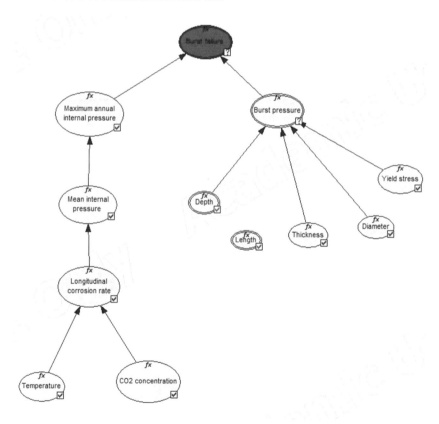

Fig. 3.1 The causality diagram to analyze the reliability of understudy energy infrastructure under material degradation corrosion form

throughout the understudy energy infrastructure. In addition, the following assumptions have been made for simulation purposes: the value of pressure and temperature are constant by time, and there is some operational pressure reduction.

Moreover, the three data sets represent the same actual pressure and temperature generated from the prior probability distributions considering the corresponding variables. The corrosion parameters are then computed based on the summation of previous values in the last time slice. Finally, the material degradation rate is approximated using the theoretical corrosion model in the literature. In this model, the rate is then multiplied by the random variables in the simulation of each trial.

The relevant MATLAB code is further developed to perform the inference on the dynamic Bayesian network. The written code could discretize every single continuous variable based on the definitions that were explained earlier. Once the variables are correctly defined, the MATLAB tool can infer using the previously discretized variables. The MATLAB tool can then conduct the elimination algorithm to filter the variables.

For the first round of the simulation set, the results indicate similar values for what is predicted from the semi-empirical corrosion model in the existing literature. The degradation rate is determined in an interval of 0.9–1.3 in the second round, highlighting the lower degradation rate, the range of the factor varies between 0.60 and 1.00. Finally, the third round of the simulation provides a higher prediction value, varying from 1.1 to 1.45. The approximated material degradation characteristics are depicted in Fig. 3.2. It should be noted that the depth of infrastructure reaches a limited state function when it reaches and passes the thickness of the infrastructure wall.

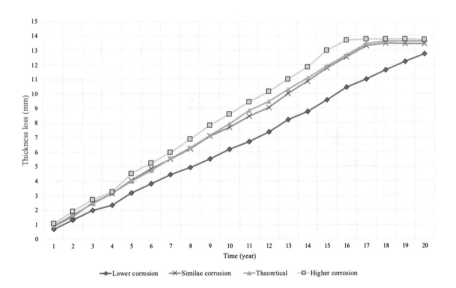

Fig. 3.2 The simulated corrosion depth values

Moreover, in the last simulation, it is assumed that the temperature remains constant; however, operational pressure declines by almost 23% every five years. Thus, this corresponds to the final phase of the energy infrastructure life cycle. The pressure and temperature simulation are utilized to compute the rate of material degradation and corrosion parameters (i.e., depth and length). Then, the corrosion parameters are computed considering the same approach, multiplying by a factor in an interval of 0.9–1.3.

Moreover, in Fig. 3.3, the reliability index is illustrated as a time function with a comparison between the values derived from the simulation with and without observed evidence. The comparison is carried out between the semi-empirical corrosion model and one observed evidence. In this case, code B31G is applied to estimate the value of burst pressure. The input data is simulated with values similar to the anticipated ones.

As can be seen from Fig. 3.3, there is a negligible increase in the reliability index based on the observed data. This is because of the sharp decrease in data uncertainty used in the developed dynamic Bayesian network (see Fig. 3.4). In addition, Fig. 3.5 depicts the mean values of the limit state function are equal with and without considering the new evidence. The standard deviation is considerably lower than when we used new evidence as observed data (see Fig. 3.6).

It should be added that such circumstance occurs when the standard deviation values of models are not less than prior probability distributions since it happens with temperature and pressure variables. In the developed model and corresponding MATLAB code, the algorithms could combine different sources of input information, and the resulting probability distribution has less standard deviation compared to both the prior distribution and the inspected-based model.

Finally, in the case of a reliability index increase, the differentiation between the outcomes estimated with and without new evidence reduces. Because the probability distribution tale is less important than failure probability determination since the

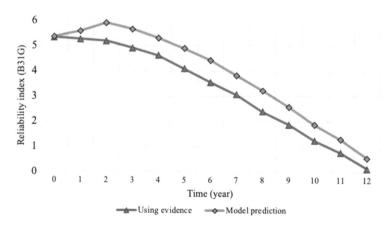

Fig. 3.3 The value of the reliability index with and without evidence

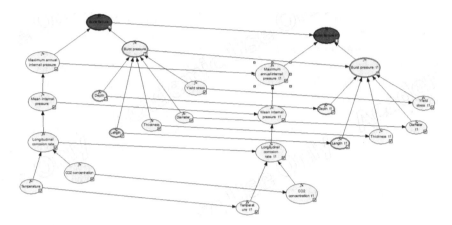

Fig. 3.4 The causality diagram to analyze the reliability of understudy energy infrastructure under material degradation corrosion form (dynamic Bayesian network in two-time slices)

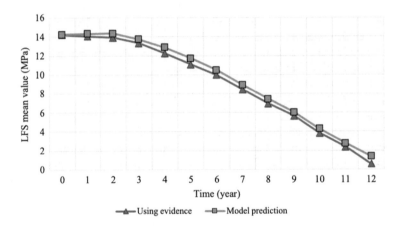

Fig. 3.5 The limit state function and mean values

probability curves are in the failure domains. However, it is vital to mention that these failures have not even happened in real-case energy infrastructure reliability analysis.

As shown in Fig. 3.7, considering the burst pressure limit state function, the lower reliability index occurs then the material degradation rate reaches half of 50% energy infrastructures thickness loss.

In a Bayesian network, the nodes represent variables, and the edges represent the dependencies between them. Each node represents a random variable, which can take on one of a set of possible values. The edges represent the conditional probabilities of one variable given its parent variables. The conditional probabilities are specified as each node's dependent probability table (CPT).

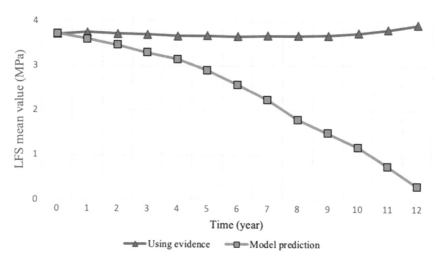

Fig. 3.6 The limit state function and standard deviation

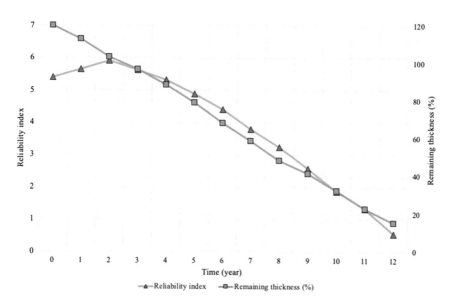

Fig. 3.7 The reliability index and the remaining energy infrastructures thickness loss

Bayesian networks can be used for various tasks, including classification, prediction, diagnosis, decision-making, and more. They have applications in many fields, including artificial intelligence, machine learning, statistics, and decision theory. One of the benefits of using a Bayesian network is that it allows for probabilistic inference. Given some evidence about the state of one or more variables, a Bayesian network can be used to calculate the probability distribution over the remaining variables [32,

33]. This can be useful for predicting outcomes, diagnosing medical conditions, or making decisions under uncertainty.

There are several advantages to using Bayesian networks, including:

1. Uncertainty modelling: Bayesian networks are designed to model uncertainty and make decisions in the presence of incomplete or uncertain information. They can be used to represent and reason about uncertain events or situations, which is a common feature of many real-world problems.
2. Interpretability: Bayesian networks provide a graphical representation of the relationship between variables, making them easy to understand and interpret. The model's visual representation can help stakeholders gain insight into the decision-making process and the factors that contribute to the outcome.
3. Predictive modelling: Bayesian networks can be used to model and predict the probability of an event occurring based on the values of the variables in the network. This can be used to identify the most likely outcome given specific inputs and to explore "what-if" scenarios to understand the impact of different variables.
4. Learning from data: Bayesian networks can be learned from data, making them a valuable tool for data analysis and machine learning. They can be used to discover patterns and relationships in data and predict outcomes based on historical data.
5. Decision-making: Bayesian networks can make decisions based on available evidence. They can be used to calculate the expected value of different actions, considering the uncertainty and risks associated with each option.

Overall, Bayesian networks are a powerful tool for modelling and reasoning under uncertainty, making them valuable in many fields [34, 35].

3.4 Conclusion

The present study demonstrated that a dynamic Bayesian modeling network has a high capacity to analyze the reliability of critical offshore infrastructure under the influence of material degradation and improve its operational performance over time. A simple corrosion type as a material deterioration is considered for assessment purposes. It has been shown that the reliability of critical offshore has increased in case of adding more observed data and pieces of evidence. This is the fact that the utilized Bayesian approach can adequately deal with both model and data uncertainty. According to this point, the accuracy in corrosion estimation does not guarantee safe operations; but a more assessment of different operation management activities can do. The results of the present study can be used to develop a risk-based critical offshore infrastructure integrity management plan, ensuring operational safety and minimizing the cost of management practices.

The dynamic Bayesian modeling network is a robust and powerful tool that automatically updates the results over system safety functionalities. In addition, it is also a flexible tool that can be utilized in other types of degradation processes and other application problems in much more complex critical systems.

References

1. Yazdi, M., Khan, F., Abbassi, R., Quddus, N., Castaneda-Lopez, H.: A review of risk-based decision-making models for microbiologically influenced corrosion (MIC) in offshore pipelines. Reliab. Eng. Syst. Saf. 108474 (2022). https://doi.org/10.1016/j.ress.2022.108474
2. Adumene, S., Khan, F., Adedigba, S.: Operational safety assessment of offshore pipeline with multiple MIC defects. Comput. Chem. Eng. 106819 (2020). https://doi.org/10.1016/j.compchemeng.2020.106819
3. Yazdi, M., Khan, F., Abbassi, R.: Operational subsea pipeline assessment affected by multiple defects of microbiologically influenced corrosion. Process Saf. Environ. Prot. **158**, 159–171 (2021). https://doi.org/10.1016/j.psep.2021.11.032
4. Hasan, S., Khan, F., Kenny, S.: Probability assessment of burst limit state due to internal corrosion. Int. J. Press. Vessel. Pip. **89**, 48–58 (2012). https://doi.org/10.1016/j.ijpvp.2011.09.005
5. Aljaroudi, A., Khan, F., Akinturk, A., Haddara, M., Thodi, P.: Risk assessment of offshore crude oil pipeline failure. J. Loss Prev. Process Ind. **37**, 101–109 (2015). https://doi.org/10.1016/j.jlp.2015.07.004
6. Leira, B.J., Næss, A., Brandrud Næss, O.E.: Reliability analysis of corroding pipelines by enhanced Monte Carlo simulation. Int. J. Press. Vessel. Pip. **144**, 11–17 (2016). https://doi.org/10.1016/j.ijpvp.2016.04.003
7. Oliveira, N., Bisaggio, H., Netto, T.: Probabilistic Analysis of the Collapse Pressure of Corroded Pipelines (2016). https://doi.org/10.1115/OMAE2016-54299
8. Gerginov, E., Rathbone, A., Sullivan, C., Griffiths, T.: Insights in the Application of Structural Reliability Analysis (SRA) for Challenging Pipeline Lateral Buckling Design (2014). https://doi.org/10.1115/OMAE2014-24450
9. Shekari, E., Khan, F., Ahmed, S.: A predictive approach to fitness-for-service assessment of pitting corrosion. Int. J. Press. Vessel. Pip. **137**, 13–21 (2015). https://doi.org/10.1016/j.ijpvp.2015.04.014
10. Xu, B., Chen, D., Behrens, P., Ye, W., Guo, P., Luo, X.: Modeling oscillation modal interaction in a hydroelectric generating system. Energy Convers. Manag. **174**, 208–217 (2018). https://doi.org/10.1016/j.enconman.2018.08.034
11. Li, H., Yazdi, M.: Advanced Decision-Making Methods and Applications in System Safety and Reliability Problems. Springer, Cham (2022). https://link.springer.com/book/9783031074295
12. Adumene, S., Okwu, M., Yazdi, M., Afenyo, M., Islam, R., Orji, C.U., Obeng, F., Goerlandt, F.: Dynamic logistics disruption risk model for offshore supply vessel operations in Arctic waters. Marit. Transp. Res. **2**, 100039 (2021). https://doi.org/10.1016/j.martra.2021.100039
13. Li, H., Yazdi, M., Huang, H.-Z., Huang, C.-G., Peng, W., Nedjati, A., Adesina, K.A.: A fuzzy rough copula Bayesian network model for solving complex hospital service quality assessment. Complex Intell. Syst. (2023). https://doi.org/10.1007/s40747-023-01002-w.
14. Yazdi, M., Adumene, S., Zarei, E.: Introducing a probabilistic-based hybrid model (Fuzzy-BWM-Bayesian Network) to assess the quality index of a medical service BT. In: Yazdi, M. (ed.), Linguistic Methods Under Fuzzy Information in System Safety and Reliability Analysis. Springer International Publishing, Cham, pp. 171–183 (2022). https://doi.org/10.1007/978-3-030-93352-4_8
15. Li, H., Yazdi, M.: Integration of the Bayesian network approach and interval type-2 fuzzy sets for developing sustainable hydrogen storage technology in large metropolitan areas BT. In: Li, H., Yazdi, M. (eds.), Advanced Decision-Making Methods and Applications in System Safety

and Reliability Problem. Springer International Publishing, Cham, pp. 69–85 (2022). https://doi.org/10.1007/978-3-031-07430-1_5

16. Yazdi, M., Khan, F., Abbassi, R., Quddus, N.: Resilience assessment of a subsea pipeline using dynamic Bayesian network. J. Pipeline Sci. Eng. **2**, 100053 (2022). https://doi.org/10.1016/j.jpse.2022.100053.

17. Adumene, S., Adedigba, S., Khan, F., Zendehboudi, S.: An integrated dynamic failure assessment model for offshore components under microbiologically influenced corrosion. Ocean Eng. **218**, 108082 (2020). https://doi.org/10.1016/j.oceaneng.2020.108082

18. Yazdi, M., Khan, F., Abbassi, R.: Microbiologically influenced corrosion (MIC) management using Bayesian inference. Ocean Eng. (2021). https://doi.org/10.1016/j.oceaneng.2021.108852

19. Palencia, O.G., Teixeira, A.P., Soares, C.G.: Safety of pipelines subjected to deterioration processes modeled through dynamic Bayesian networks. Int. J. Press. Vessel. Pip. **141**, 1–11 (2019). https://doi.org/10.1115/1.4040573

20. Dong, Y., Frangopol, D.M.: Risk-informed life-cycle optimum inspection and maintenance of ship structures considering corrosion and fatigue. Ocean Eng. **101**, 161–171 (2015). https://doi.org/10.1016/j.oceaneng.2015.04.020

21. Nesic, S., Nyborg, R., Stangeland, A., Nordsveen, M.: Mechanistic modeling for CO_2 corrosion with protective iron carbonate films. In: Corrosion 2001 (2001) NACE-01040

22. Pots, B.F.M.: Prediction of corrosion rates of the main corrosion mechanisms in upstream applications. In: Corrosion 2005. (2005) NACE-05550

23. de Waard, C., Lotz, U., Milliams, D.E.: Predictive model for CO_2 corrosion engineering in wet natural gas pipelines. Corrosion **47**, 976–985 (1991). https://doi.org/10.5006/1.3585212

24. Weber, P., Medina-Oliva, G., Simon, C., Iung, B.: Overview on Bayesian networks applications for dependability, risk analysis and maintenance areas. Eng. Appl. Artif. Intell. **25**, 671–682 (2012). https://doi.org/10.1016/j.engappai.2010.06.002

25. Yazdi, M., Kabir, S.: A fuzzy Bayesian network approach for risk analysis in process industries. Process Saf. Environ. Prot. **111**, 507–519 (2017). https://doi.org/10.1016/j.psep.2017.08.015

26. Fenton, N.E., Neil, M.: Risk Assessment and Decision Analysis with Bayesian Networks (2013)

27. Langseth, H., Nielsen, T.D., Rumí, R., Salmerón, A.: Inference in hybrid Bayesian networks. Reliab. Eng. Syst. Saf. **94**, 1499–1509 (2009). https://doi.org/10.1016/j.ress.2009.02.027

28. Yazdi, M., Khan, F., Abbassi, R., Rusli, R.: Improved DEMATEL methodology for effective safety management decision-making. Saf. Sci. **127**, 104705 (2020). https://doi.org/10.1016/j.ssci.2020.104705

29. Kelly, D.L., Smith, C.L.: Bayesian inference in probabilistic risk assessment-The current state of the art. Reliab. Eng. Syst. Saf. **94**, 628–643 (2009). https://doi.org/10.1016/j.ress.2008.07.002

30. Zwirglmaier, K., Straub, D.: A discretization procedure for rare events in Bayesian networks. Reliab. Eng. Syst. Saf. **153**, 96–109 (2016). https://doi.org/10.1016/j.ress.2016.04.008

31. Teixeira, A.P., Soares, C.G., Netto, T.A., Estefen, S.F.: Reliability of pipelines with corrosion defects. Int. J. Press. Vessel. Pip. **85**(85), 228–237 (2008). https://doi.org/10.1016/j.ijpvp.2007.09.002

32. Nedjati, A., Yazdi, M., Abbassi, R.: A sustainable perspective of optimal site selection of giant air-purifiers in large metropolitan areas. Environ. Dev. Sustain. **24**, 8747–8778 (2022)

33. OpenAI, ChatGPT [Software] (2021). https://openai.com

34. Yazdi, M., Mohammadpour, J., Li, H., Huang, H.-Z., Zarei, E., Pirbalouti, R.G., Adumene, S.: Fault tree analysis improvements: a bibliometric analysis and literature review. Qual. Reliab. Eng. Int. n/a (2023). https://doi.org/10.1002/qre.3271

35. Yazdi, M.: A review paper to examine the validity of Bayesian network to build rational consensus in subjective probabilistic failure analysis. Int. J. Syst. Assur. Eng. Manag. **10**, 1–18 (2019). https://doi.org/10.1007/s13198-018-00757-7

Chapter 4
An Improved LeNet-5 Convolutional Neural Network Supporting Condition-Based Maintenance and Fault Diagnosis of Bearings

Abstract This chapter introduces an improved LeNet-5 convolutional neural network model for condition-based maintenance and fault diagnosis of bearings. The model can effectively extract the fault characteristics of one-dimensional vibration signals. To be specific, time–frequency transformation methods, short-time Fourier transform, and wavelet transform are used to convert the vibration signal into a two-dimensional time-domain-frequency domain signal so as to extend the amount of time-domain frequency domain information of the vibration signal. An improved LeNet-5 convolutional neural network model is then constructed to improve the effectiveness of the traditional model when applied to fault diagnosis of bearings. A case study based on the bearing fault dataset is implemented to validate the improved method's effectiveness, accuracy, and applicability. Overall, the proposed method contributes to the condition-based maintenance of bearings.

Keywords Condition-based maintenance · Fault diagnosis · Bearings · Improved LeNet-5 convolutional neural network

4.1 Introduction

Beadings are crucial elements of rotating machinery [1]. Condition-based maintenance of rotating machinery relies heavily on accurate fault diagnosis of bearings [2–4]. However, with the progress of industrialization, the working conditions of rotating machineries like power plants and wind turbines are becoming complex [5, 6]. Signal analysis-based fault diagnosis is reliable for fault analysis, healthy state monitoring, and condition-based maintenance [1]. The impact characteristics caused by rolling bearing faults are often submerged in the intense noise background signal [7]. Hence, accurate fault awareness, identification, diagnosis, and prediction are challenging.

The development and application of machine learning methodologies to fault diagnosis of bearings and rotational machinery have irritated the condition-based maintenance of such elements and systems, which include, but are not limited to, artificial neural networks (ANN) [8], convolutional neural networks (CNN) [9], support vector machines (SVM) [10], constrained Boltzmann machines (RBM) [11], sparse filtering

© The Author(s), under exclusive license to Springer Nature Switzerland AG 2023
H. Li et al., *Intelligent Reliability and Maintainability of Energy Infrastructure Assets*,
Studies in Systems, Decision and Control 473,
https://doi.org/10.1007/978-3-031-29962-9_4

[12], K-nearest neighbors [13], and so on but not limited to [1, 14–16] Among the aforementioned, CNN is widely applied to process two-dimensional image data and has shown advantages and promising potential in feature extraction and classification [1, 8].

In engineering, signal process technologies are used to analyze vibration signals to obtain input data of CNN [17]. Then, a corresponding CNN is constructed according to the input data to extract the fault characteristics to support fault diagnosis of rolling bearings. However, data is always insufficient to train the constructed CNN model [1, 17]. To this end, researchers devoted themselves to extending the scale of input data of CNN. For instance, Kumar et al. [18] used wavelet synchronous compression transformation (WSST) to convert one-dimensional time-series vibration data into two-dimensional images. In the study, a fault diagnosis model according to the frequency domain signals characteristics is constructed, proving to be sufficient in excavating deep abstract features in bearing signals. Zhou et al. [19] combined the vibration signal data collected by multiple sensors into a data image and constructed a rolling bearing fault diagnosis model combining transfer learning theory and a convolutional neural network. The study displays a new way of input data processing for fault diagnosis. Wang et al. [20] pointed out that: (i) the actual operating condition of bearings is complex; (ii) Fault characteristics of beadings follow different distributions; (iii) The convolution kernel of the convolutional neural network is tiny; (iv) The overall trend of the feature distribution is usually ignored. Accordingly, they designed a one-dimensional visual convolutional neural network (VCN), which has better stability than traditional networks when applied to fault diagnosis engineering systems.

Other studies on the same topic include:

- Bearing fault diagnosis method based on a convolutional autoencoder (Kaji et al. [21]).
- Bearing fault diagnosis model, namely TLCNN, based on the transfer learning theory (Ma et al. [22]).
- Data augmentation technology to create effective training samples to train the model artificially (Li et al. [23]).
- Bearing fault model combining deep convolutional network and random forest method (Xu et al. [24]).
- Normalized convolutional neural network model to solve limited data situations (Zhao et al. [25]).
- The deep learning model combined a convolutional neural network with automatic hyperparameter optimization (Li et al. [26]).

However, the limitations of the mentioned method are obvious, including (i) The methods are mainly developed for data samples without noise interference and cannot handle the problem of the vibration signal of rolling bearings mixing with noise; (ii) For the convolutional neural network to extract the corresponding features from the original data, each layer of the network structure requires a large number of neurons, which increases the model's training time.

According to the above, this chapter builds an improved LeNet-5 convolutional neural network model. The advantage of the improved LeNet-5 convolutional neural network model are as follows: (i) Based on short-time Fourier transform and wavelet transform methodologies, the time–frequency domain information of vibration signals based on one-dimensional time series is transformed into two-dimensional time–frequency domain signal; (ii) Simulate the noise condition of industrial systems by adding different proportions of Gaussian white noise to improve the ability of the model to learn compelling features, which can enhance the generalization performance of the model.

The purpose of this chapter is to introduce the improved LeNet-5 convolutional neural network model and show the application of the model to readers. The comparison with the existing methods illustrates the advantage of the model. The rest of this chapter is arranged as follows. Section 4.2 introduces the theoretical foundations. Section 4.3 describes the framework of the proposed method. Section 4.4 is case studies and discussions. Conclusions are provided in Sect. 4.5.

4.2 Preliminary

4.2.1 Convolutional Neural Network (CNN)

CNN consists of input, convolutional, pooling, fully connected, and output layers [27]. A Convolutional Neural Network (CNN) is a type of artificial neural network that is particularly well-suited for image recognition and computer vision tasks. It is inspired by the organization of the visual cortex in animals, which has a hierarchical structure specialized in processing visual information. In a CNN, the input image is fed through convolutional layers, which apply filters to the input image. The filters are typically small, square matrices of values learned during the training process. The convolution operation applies each filter to a local region of the input image, producing a feature map highlighting certain features in the input image, such as edges or textures. After the convolutional layers, the feature maps are fed through a series of pooling layers. These downsample the feature maps by taking the maximum or average value over a tiny map region. This reduces the feature maps' size and helps make the network more computationally efficient. The convolutional layer and the pooling layer constitute the convolution module, and the deep CNN consists of multiple convolutional blocks stacked to extract the features of the input data layer by layer [27, 28]. The structure of CNN is shown in Fig. 4.1.

The convolutional layer is to extract features received by the input layer [29]. In image recognition, the input image is high-dimensional, making each neuron fully connected with all neurons, which is a problem for training CNNs. Therefore, the convolutional layer extracts local features through multiple local perception modules, and the corresponding numerous feature maps can be obtained through various convolution kernel calculations [30]. Figure 4.2 shows the convolution process using a

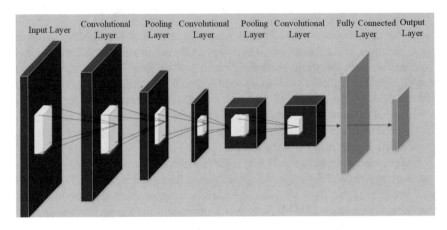

Fig. 4.1 Structure of CNN

convolution kernel of size 3×3 on an input matrix of 5×5. The convolution operation is shown in Eq. (4.1).

$$X = f\left(\sum x * w + b\right) \tag{4.1}$$

where, $*$ indicates the convolution operator, x is the input image, w is the convolution kernel, b represents the bias, and $f(\cdot)$ demotes the activation function.

A pooling layer locates between two convolutional layers to reduce the scale of parameters and the overfitting problem of neural networks [31]. The pooling layer reduces the model's complexity by reducing the feature image's size [29]. It compresses the feature information in the image so that the model extracts the main features. The pooling consists of maximum pooling, average pooling, and weighted pooling [32–35]. The most commonly used pooling method is maximum pooling; see Eq. (4.2) and Fig. 4.3.

$$P_{ijk}^{l} = \max(y_{ijk}^{l} : i \le i < i + p, j \le j < j + p) \tag{4.2}$$

where P represents the length of the pooling window, q represents the width.

The convolution module processes the input data and sends the processed data to the fully connected layer. The operation of the fully connected layer is as in Eq. (4.4).

$$y^{k} = f\left(w^{k} \times x^{k-1} + b^{k}\right) \tag{4.4}$$

where, k represents the ordinal number of the network layer, y^{k} reflects the output of the fully connected layer, x^{k-1} is the feature vector, w^{k} is the weight coefficient, and represents the bias.

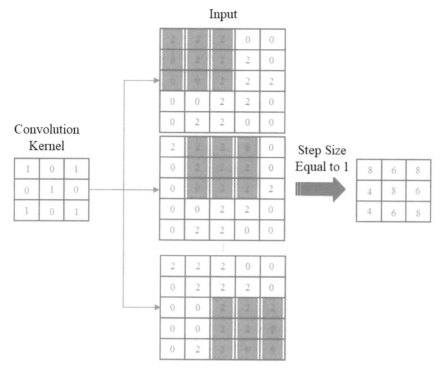

Fig. 4.2 The convolution operation

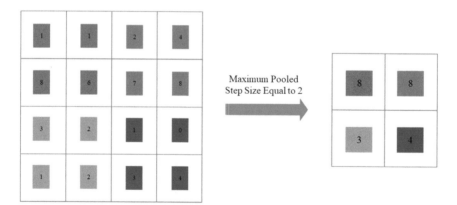

Fig. 4.3 The maximum pooled sampling

4.2.2 Time–Frequency Analysis of Vibration Signal

Time–frequency analysis methods have been applied to signal characteristics analysis [36]. The information of the signal in the time domain and frequency domain can be extracted by time–frequency analysis. This section introduces the short-time Fourier transform and wavelet transform methods.

The short-term Fourier transform is one of the most commonly used time–frequency analysis methods, see Eq. (4.5), which uses a window function with a selectable length to translate the number of selectable overlapping points on time domain signals [37]. Fourier transforms, splicing and recombining each segment, and finally obtains a two-dimensional time–frequency spectrum [37, 38].

$$STFT_x(\tau, f) = \int_{-\infty}^{+\infty} x(t)h(t - \tau)e^{-j2\pi ft} dt \qquad (4.5)$$

where, $x(t)$ is the original time domain signal, $h(t - \tau)e^{-j2\pi ft}$ is the primary function, f and is the frequency of the Fourier transform.

The wavelet transform is window size adaptable, which automatically adjusts size based on the time domain characteristics of signals [39]. The window size can be compressed to obtain a higher time resolution in case the signal frequency is high.

The principle of wavelet transform is as follows:

Let $\varphi(t) \in L^2(R)$ be the square-integrable real space, where $\phi(\omega)$ is a function $\varphi(t)$ obtained by the Fourier transform if the Formula (4.6) is satisfied:

$$0 < C_\phi = \int_{-\infty}^{+\infty} \frac{|\phi(\omega)|^2}{|\omega|} d\omega < \infty \qquad (4.6)$$

So that $\varphi(t)$ is called the parent wavelet function, and different wavelet functions can be obtained by manipulating the coefficients of the parent wavelet function, see Eq. (4.7).

$$\varphi_{a,b}(t) = \frac{1}{\sqrt{a}}\varphi\left(\frac{t - b}{a}\right) \quad a, b \in R, a \neq 0 \qquad (4.7)$$

where, a is the scale factor; b is the translation factor.

Wavelet transforms operate the signal $x(t)$ according to the following:

$$0 < C_\phi = \int_{-\infty}^{+\infty} \frac{|\phi(\omega)|^2}{|\omega|} d\omega < \infty CWT(a, b) = \frac{1}{\sqrt{a}} \int_{-\infty}^{+\infty} x(t)\varphi\left(\frac{t - b}{a}\right) dt \qquad (4.8)$$

4.3 The Proposed Model for Bearing Fault Diagnosis

The LeNet-5 network model is a basic convolutional neural network [40], the structure of which is shown in Fig. 4.4.

The input layer of the LeNet-5 model accepts two-dimensional images with input data in size 32 × 32. The first convolutional layer contains six convolution kernels with a size of 5 × 5. After passing the first layer convolution operation, a featured image (28 × 28) can be obtained in the output. The convolution operation is shown in Eq. (4.9).

$$x_j^i = \int \left(\sum_{i \in M_j} x_i^{l-1} * k_{ij}^l + b_j^l \right) \tag{4.9}$$

where, l indicates the number of layers; k represents the convolution kernel; M_j represents the feature map of j; b is the bias; $\int$ denotes the activation function of the convolutional layer.

The pooling module of the first pooling layer is a 2 × 2 matrix, the pooling stride is 2, and the output data after the pooling operation is a 6-feature map with the size 14 × 14, see Eq. (4.10).

$$x_j^i = f \left(\beta_j^l f_\bullet \left(x_j^{l-1} + b_j^l \right) \right) \tag{4.10}$$

where, $f_\bullet$ represents the pooling function β, and b are feature parameters.

The second convolutional layer consists of 16 convolution kernels with a size of 5 × 5. The convolution operation of the second convolutional layer exports a featured image with a data size of 10 × 10. After the operation of the second pooling layer, the output data size is 16. The third convolutional layer consists of 120 convolution

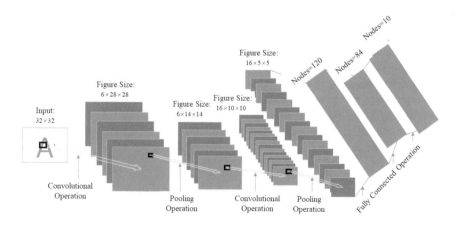

Fig. 4.4 The structure of the LeNet-5 model

kernels with a size of 5×5. After the convolution operation of the third convolutional layer, the output data size is 11. The fully connected layer consists of two layers; the final output is 10 neurons.

The traditional LeNet-5 has limitations in the convergence speed of model training and generalization. This section proposes some improvements to the traditional LeNet-5 with the following modifications:

- The input layer of the traditional LeNet-5 network accepts two-dimensional images. In practice, the larger the two-dimensional image, the more accurate the fault diagnosis of the model. Therefore, the input layer acceptance data of the LeNet-5 network is extended to a 64×64-pixel two-dimensional time–frequency image to contain more feature information and increase the training efficiency of the model.
- The feature extraction ability of the network model is improved by increasing the convolution kernels of each convolutional layer. Specifically, the convolution kernel of the first convolutional layer of the traditional LeNet-5 network was improved from six convolutional kernels of 5×5 to eight convolution kernels of 8×8.
- The batch normalization method is introduced into the model to improve the model's training efficiency and enhance the model's ability to suppress overfitting.

Table 4.1 shows the specific parameters of the improved LeNet-5 network. The steps to implement the improved LeNet-5 network are as follows:

Step 1: Collect vibration signals of bearings under different working states by sensors.

Step 2: Divide the dataset collected into training and test sets. The corresponding labels are made according to the different fault categories of bearings. The one-dimensional data with labels are subjected to wavelet transform and short-time Fourier transform, and the corresponding time–frequency diagrams can be obtained.

Step 3: Initialize the parameters of the LeNet-5 network (see Table 4.1). According to the size of the time–frequency graph, determine the number of nodes of the model and initialize the weights and biases of the network.

Step 4: Import the time–frequency graph data into the LeNet-5 network. The labels corresponding to the fault type are obtained through the full connection and classification layers. The error is calculated using the backpropagation algorithm, updating the weight and bias.

Step 5: Import the features extracted by the improved LeNet-5 network and the corresponding training labels into the softmax classifier.

The framework of bearing fault diagnosis based on the improved LeNet-5 network model is shown in Fig. 4.5.

Layer name	Parameter	Value
Table 4.1 Detailed parameters of the improved LeNet-5 network		
Convolutional layer 1	Convolution kernel size	8×8
	The number of kernels	8
	Number of channels 1	Stride 1
Pooling layer 1	Pooling mode	Max pooling
	Pool size	2×2
	Stride	2
Convolutional layer 2	Convolution kernel size	8×8
	The number of kernels	32
	Number of channels 1	Stride 1
Pooling layer 2	Pooling mode	Max pooling
	Pool size	2×2
	Stride	2
Convolutional layer 3	Convolution kernel size	5×5
	The number of kernels	64
	Number of channels 1	Stride 1
Pooling layer 3	Pooling mode	Max pooling
	Pool size	2×2
	Stride	2
Fully connected layer 1	Number of neurons	512
Fully connected layer 2	Number of neurons	128
Fully connected layer 3 (Output layer)	Number of neurons	10
	Classifier	Softmax

4.4 Case Study and Validation

4.4.1 Data Preprocessing

This chapter uses the bearing dataset published by Case Western Reserve University [41] to train and test the performance of the proposed model. The practical device consists of a power tester, a torque sensor, an acceleration sensor, and an electric motor. And Electrical Discharge Machining (EDM) technology forms single-point damage on bearings to simulate failure conditions. The failure severity of bearings is simulated by the size of the EDM bearing diameter, where the diameter of the EDM bearing includes 0.18 mm (millimeter), 0.36 mm, and 0.54 mm. The fault locates at the inner ring, outer ring, and rolling element.

In this chapter, the vibration data of nine different fault states of the drive end and that of the normal state are studied. The sampling frequency of the bearing vibration signal is 48 kHz, the speed of the motor is 1724 r/min, and 1500 data points can be collected in one cycle of rotation. The data points collected by one-third of the

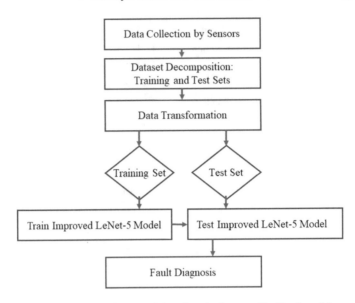

Fig. 4.5 Bearing fault diagnosis framework based on the improved LeNet-5 model

rotation cycle of the motor shaft are taken as a sample. Overall, 500 consecutive data points are selected as a sample, 800 samples are collected for each fault type as the training set, and another 150 samples are selected as the test set. The total number of training samples in the dataset is 8000, and the total number of test set samples is 150. The fault types and fault labels for rolling bearings are shown in Table 4.2.

Bearing failure data is based on a one-dimensional vibration signal in time series. The short-time Fourier transform time spectrum generates the vibration data of different states. According to the rolling bearing failure data division, 10 label data types are subjected to a short-term Fourier transform. Figure 4.6 shows short-term

Table 4.2 Details of fault data of bearings

Fault location	Diameter (mm)	Label
Normal	0	1
Ball	0.18	2
Ball	0.36	3
Ball	0.54	4
Inner	0.18	5
Inner	0.36	6
Inner	0.54	7
Outer	0.18	8
Inner	0.36	9
Inner	0.54	10

Fourier transform time–frequency diagrams of the four different states of rolling bearings: inner ring fault, rolling element fault, outer ring fault, and normal.

The wavelet transformation of bearing vibration data in different states generates a time–frequency diagram. According to the above division of rolling bearing fault data, the wavelet transformation of 10 label data types is carried out. The wavelet function adopts the CMOR1-3 wavelet basis function to obtain the corresponding time–frequency diagram. Figure 4.7 shows the time–frequency plot obtained by wavelet transform for ten fault conditions.

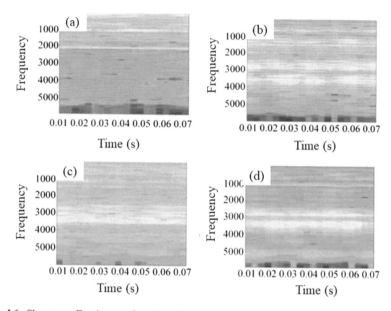

Fig. 4.6 Short-term Fourier transform time–frequency diagrams **a** normal state; **b** inner ring fault; **c** rolling element fault; **d** outer ring fault

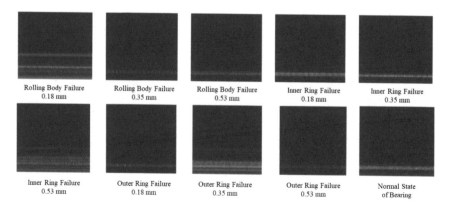

Fig. 4.7 Wavelet time–frequency diagrams under ten conditions

The two-dimensional time–frequency image set obtained by short-term Fourier transform based on one-dimensional time-series vibration signals in the previous article is defined as dataset 1 (with a time–frequency graph size of 64 × 64); The two-dimensional time–frequency image obtained by wavelet transformation based on one-dimensional time-series vibration signals is defined as dataset 2 (64 × 64).

4.4.2 Results

The simulation experiment adopts the pre-training mode. The mean square error (MSE) is used as the loss function. The Adams optimizer is employed to update the weight and bias of the model. The effectiveness of the fault diagnosis model is verified by using the dataset set in Sect. 4.1. It contains dataset 1, generated by the short-time Fourier transform, and dataset 2, generated by the wavelet transform. To eliminate the influence of model randomness, the simulation was repeated ten times, and the fault diagnosis accuracies of the model using datasets 1 and 2 are shown in Fig. 4.8. Table 4.3 lists the average accuracy and standard deviation of fault diagnosis.

It can be seen from Table 4.3 that after 10 repeated simulations, the average accuracy of fault diagnosis based on the improved LeNet-5 network and dataset 1

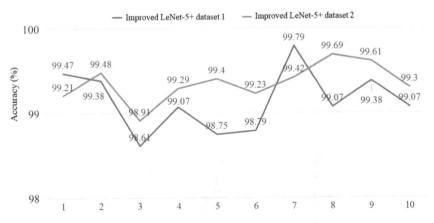

Fig. 4.8 Fault diagnosis accuracy on data sets 1 and 2 under various conditions

Table 4.3 Average accuracy and standard deviation of fault diagnosis	Model	Average accuracy (%)	Standard deviation
	Improved LeNet-5 + dataset1	99.13	0.367
	Improved LeNet-5 + dataset2	99.35	0.221

is 99.13%, and the same parameter based on dataset 2 is 99.35%. It indicates that the proposed method achieves high accuracy of fault identification based on the datasets obtained by the two time–frequency transformation methods. The average fault diagnosis accuracy based on dataset 2 obtained by wavelet transform is higher than dataset 1 obtained by short-time Fourier transform.

Figure 4.9 displays the fault diagnosis classification confusion matrix of the improved LeNet-5 structure based on dataset 1, which is more than 99% under the condition category labels 1, 2, 4, 7, 8, 9, and 10. Still, the accuracy of the working condition category labels 3, 5, and 6 is relatively lower. The accuracy rate of the working condition category 5 is the lowest. The same indicators regarding dataset 2 are listed in Fig. 4.10 and the diagnostic accuracy of each working condition is more than 99%. According to Fig. 4.10, the diagnostic accuracy of category 3 is the lowest.

Real Label	1	2	3	4	5	6	7	8	9	10
1	0.9947	0.0000	0.0000	0.0000	0.0017	0.0000	0.0000	0.0000	0.0000	0.0000
2	0.0000	0.9938	0.0022	0.0000	0.0020	0.0000	0.0000	0.0000	0.0000	0.0000
3	0.0000	0.0069	0.9861	0.0050	0.0000	0.0000	0.0000	0.0000	0.0000	0.0000
4	0.0000	0.0020	0.0033	0.9907	0.0000	0.0000	0.0000	0.0000	0.0000	0.0000
5	0.0000	0.0000	0.0000	0.0042	0.9875	0.0000	0.0000	0.0041	0.0000	0.0000
6	0.0000	0.0000	0.0000	0.0000	0.0020	0.9879	0.0020	0.0000	0.0000	0.0000
7	0.0000	0.0000	0.0000	0.0000	0.0000	0.0020	0.9979	0.0000	0.0036	0.0000
8	0.0000	0.0000	0.0000	0.0000	0.0000	0.0042	0.0000	0.9907	0.0000	0.0000
9	0.0000	0.0000	0.0000	0.0000	0.0000	0.0000	0.0000	0.0000	0.9938	0.0020
10	0.0000	0.0000	0.0000	0.0000	0.0000	0.0020	0.0000	0.0000	0.0000	0.9907

Predicted Label

Fig. 4.9 Classification confusion matrix of the improved LeNet-5 model based on short-time Fourier transform

Real Label	1	2	3	4	5	6	7	8	9	10
1	0.9921	0.0000	0.0000	0.0000	0.0017	0.0000	0.0000	0.0000	0.0000	0.0000
2	0.0000	0.9948	0.0022	0.0000	0.0020	0.0000	0.0000	0.0000	0.0000	0.0000
3	0.0000	0.0069	0.9891	0.0050	0.0000	0.0000	0.0000	0.0000	0.0000	0.0000
4	0.0000	0.0020	0.0033	0.9929	0.0000	0.0000	0.0000	0.0000	0.0000	0.0000
5	0.0000	0.0000	0.0000	0.0042	0.9940	0.0000	0.0000	0.0041	0.0000	0.0000
6	0.0000	0.0000	0.0000	0.0000	0.0020	0.9923	0.0020	0.0000	0.0000	0.0000
7	0.0000	0.0000	0.0000	0.0000	0.0000	0.0020	0.9942	0.0000	0.0036	0.0000
8	0.0000	0.0000	0.0000	0.0000	0.0000	0.0042	0.0000	0.9969	0.0000	0.0000
9	0.0000	0.0000	0.0000	0.0000	0.0000	0.0000	0.0000	0.0000	0.9961	0.0020
10	0.0000	0.0000	0.0000	0.0000	0.0000	0.0020	0.0000	0.0000	0.0000	0.9930

Predicted Label

Fig. 4.10 Classification confusion matrix of the improved LeNet-5 model based on wavelet transform

It can be seen from Figs. 4.9 and 4.10 that the LeNet-5 structure based on a short-time Fourier transform and that based on the wavelet transform process satisfied results with high classification accuracy and stable classification performance, except for a few samples.

Vibration signals of bearings are affected by noise and the resonance of the entire system, which will seriously affect the accuracy of fault diagnosis. This section verifies the fault diagnosis accuracy of the proposed improved LeNet-5 model in a noisy environment. Different scales of white Gaussian noise are added to the original one-dimensional time-series vibration signal to test the fault diagnosis accuracy of the model. The signal-to-noise ratio (SNR) is defined as shown in Eq. (4.11).

$$SNR = 10\log_{10}\left(\frac{P_{signal}}{P_{noise}}\right) \qquad (4.11)$$

where, P_{signal} is the power of the original vibration signal, p_{noise} is the noise power.

The vibration signal is modulated into a noise signal with different SNRs. Specifically, in this study, SNRs are -4 dB, -2 dB, 0 dB, 2 dB, and 4 dB, respectively. Table 4.4 shows the fault diagnosis accuracy of the improved LeNet-5 model on datasets 1 and 2 under different SNRs.

It can be seen from Table 4.4 that the improved LeNet-5 model process high accuracies of fault diagnosis for different degrees of noise. For instance:

- When SNR = 4, the noise interference degree is the smallest, and the fault diagnosis accuracy of the model based on dataset 1 pre-processed by short-time Fourier transform is 97.35%; the same performance for dataset 2 pre-processed by wavelet transform is 99.20%.
- When SNR = -4, the noise interference degree is the highest, and the fault diagnosis accuracy of the model based on datasets 1 and 2 are 69.79% and 83.70%, respectively.

However, a higher fault diagnosis accuracy is obtained when the improved wavelet transformation method is implemented in the LeNet-5 schedule. It indicates that the wavelet transformation pre-processing of fault data can effectively reduce the influence of redundant noise.

Table 4.4 Accuracy of fault diagnosis under SNRs

SNR (dB)	Fault diagnosis accuracy (%)	
	Improved LeNet-5 + dataset1	Improved LeNet-5 + dataset2
−4	69.79	83.70
−2	85.69	90.70
0	91.27	94.68
2	93.48	98.25
4	97.35	99.20

To further verify the superiority of the fault diagnosis method based on the improved LeNet-5 model, this section compares the improved LeNet-5 model proposed with other five fault diagnosis methods, including Support Vector Machine (SVM), k-Nearest Neighbor (k-NN), K-Means, Back Propagation Neural Network (BPNN), and the traditional LeNet-5 network. The parameters of the traditional LeNet-5 network are the same as those listed in Table 4.1, and the network parameters of all models are the same as in Table 4.5.

The fault diagnosis accuracy of each method is shown in Fig. 4.11. Table 4.6 shows the average accuracy of fault diagnosis of each method.

From Table 4.6, it can be seen that the diagnostic accuracy of the improved LeNet-5 network is increased when compared to other methods, for instance, it increased by 6.13% compared with SVM and also have a better performance than other methods like K-NN (11.19%), K-Means (11.68%), and BPNN (9.33%). It is noted that the diagnostic accuracy of the improved LeNet-5 network is increased by 10.5% compared to the conventional LeNet-5 network. It is known that the input data of the compared methods, such as SVM, K-NN, K-Means, and BPNN are one-dimensional vibration signals, limiting their fault diagnosis ability.

Table 4.5 Detailed network parameters

Methods	Parameter settings
SVM	$C = 1$; the radial basis function is the kernel function, and $\gamma = 0.125$
K-NN	The number of nearest neighbors is 5
K-means	The maximum number of iterations is 1000; the number of clusters is 4
BPNN	10 input layer nodes; 12 hidden layer nodes; 4 output layer nodes; learning rate is 0.003; the maximum iterations is 1000

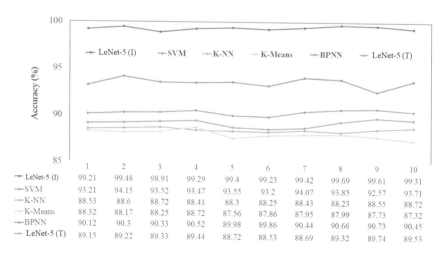

	1	2	3	4	5	6	7	8	9	10
LeNet-5 (I)	99.21	99.48	98.91	99.29	99.4	99.23	99.42	99.69	99.61	99.31
SVM	93.21	94.15	93.52	93.47	93.55	93.2	94.07	93.85	92.57	93.71
K-NN	88.53	88.6	88.72	88.41	88.3	88.25	88.43	88.23	88.55	88.72
K-Means	88.32	88.17	88.25	88.72	87.56	87.86	87.95	87.99	87.73	87.32
BPNN	90.12	90.3	90.33	90.52	89.98	89.86	90.44	90.66	90.73	90.45
LeNet-5 (T)	89.15	89.22	89.33	89.44	88.72	88.53	88.69	89.32	89.74	89.53

Fig. 4.11 Accuracy of fault diagnosis of each method/LeNet-5 (T): Traditional LeNet-5; LeNet-5 (I): Improved LeNet-5

Table 4.6 Fault diagnosis accuracy of different fault diagnosis methods

Diagnosis method	Average fault diagnosis accuracy (%)
SVM	93.53
K-NN	88.47
K-means	87.98
BPNN	90.33
Traditional LeNet-5	89.16
Proposed LeNet-5	99.66

It is also pointed out that fault diagnosis is the fundamental step of condition-based maintenance. The method proposed in this book chapter provides an effective tool for fault diagnosis of bearings which is the key element for rotational machinery and energy infrastructure assets and will support the subsequent condition-based maintenance of the mentioned systems.

4.5   Conclusion

This chapter introduces an improved LeNet-5 convolutional neural network model for condition-based maintenance and fault diagnosis of bearings. The model can effectively extract the fault characteristics of one-dimensional vibration signals based on time series. Short-time Fourier transform, and wavelet transform are used to convert the vibration signal based on the one-dimensional time series into a two-dimensional time-domain-frequency domain signal, based on which accurate fault diagnosis of bearings can be completed with the assistance of the improved LeNet-5 convolutional neural network model. A case study based on a bearing fault dataset is implemented to validate the effectiveness, accuracy, and applicability of the improved method. Overall, the reader would realize that the proposed method contributes to the condition-based maintenance of bearings of rotational machinery and energy infrastructure assets. However, as a direction for future study, it can be integrated with other types of intelligent tools into the energy infrastructure, which some have done by authors previously [42–47].

References

1. Liu, R., Yang, B., Zio, E., Chen, X.: Artificial intelligence for fault diagnosis of rotating machinery: a review. Mech. Syst. Signal Process. **108**, 33–47 (2018)
2. Wei, Y., Li, Y., Xu, M., Huang, W.: A review of early fault diagnosis approaches and their applications in rotating machinery. Entropy **21**(4), 409 (2019)
3. Yazdi, M., Mohammadpour, J., Li, H., Huang, H.-Z., Zarei, E., Pirbalouti, R.G., Adumene, S.: Fault tree analysis improvements: a bibliometric analysis and literature review. Qual. Reliab. Eng. Int. n/a (2023). https://doi.org/10.1002/qre.3271

4. Li, H., Soares, C.G., Huang, H.Z.: Reliability analysis of a floating offshore wind turbine using Bayesian networks. Ocean Eng. **217**, 107827 (2020)
5. Li, Y., Wang, X., Liu, Z., Liang, X., Si, S.: The entropy algorithm and its variants in the fault diagnosis of rotating machinery: a review. IEEE Access **6**, 66723–66741 (2018)
6. Li, H., Soares, C.G.: Assessment of failure rates and reliability of floating offshore wind turbines. Reliab. Eng. Syst. Saf. **228**, 108777 (2022)
7. Jia, F., Lei, Y., Lin, J., Zhou, X., Lu, N.: Deep neural networks: a promising tool for fault characteristic mining and intelligent diagnosis of rotating machinery with massive data. Mech. Syst. Signal Process. **72**, 303–315 (2016)
8. Karatuğ, Ç., Arslanoğlu, Y.: Development of condition-based maintenance strategy for fault diagnosis for ship engine systems. Ocean Eng. **256**, 111515 (2022)
9. Jiang, X., Yang, S., Wang, F., Xu, S., Wang, X., Cheng, X.: OrbitNet: a new CNN model for automatic fault diagnostics of turbomachines. Appl. Soft Comput. **110**, 107702 (2021)
10. Kang, J., Wang, Z., Guedes Soares, C.: Condition-based maintenance for offshore wind turbines based on support vector machine. Energies **13**(14), 3518 (2020)
11. Mocanu, D.C., Mocanu, E., Nguyen, P.H., Gibescu, M., Liotta, A.: A topological insight into restricted Boltzmann machines. Mach. Learn. **104**(2), 243–270 (2016)
12. Zhang, Z., Li, S., Wang, J., Xin, Y., An, Z.: General normalized sparse filtering: a novel unsupervised learning method for rotating machinery fault diagnosis. Mech. Syst. Signal Process. **124**, 596–612 (2019)
13. Zhou, Z., Wen, C., Yang, C.: Fault isolation based on k-nearest neighbor rule for industrial processes. IEEE Trans. Industr. Electron. **63**(4), 2578–2586 (2016)
14. Li, H., Yazdi, M.: Developing failure modes and effect analysis on offshore wind turbines using two-stage optimization probabilistic linguistic preference relations. In: Advanced Decision-Making Methods and Applications in System Safety and Reliability Problems, pp. 47–68. Springer International Publishing, Cham (2022)
15. Li, H., Yazdi, M., Huang, H.-Z., Huang, C.-G., Peng, W., Nedjati, A., Adesina, K.A.: A fuzzy rough copula Bayesian network model for solving complex hospital service quality assessment. Complex Intell. Syst. (2023). https://doi.org/10.1007/s40747-023-01002-w.
16. Li, H., Yazdi, M.: Reliability analysis of correlated failure modes by transforming fault tree model to Bayesian network: a case study of the MDS of a CNC machine tool. In: Advanced Decision-Making Methods and Applications in System Safety and Reliability Problems. Studies in Systems, Decision and Control, vol. 211. Springer, Cham.
17. Patole, S.M., Torlak, M., Wang, D., Ali, M.: Automotive radars: a review of signal processing techniques. IEEE Signal Process. Mag. **34**(2), 22–35 (2017)
18. Kumar, A., Gandhi, C.P., Zhou, Y., Vashishtha, G., Xiang, J.: Improved CNN for the diagnosis of engine defects of 2-wheeler vehicle using wavelet synchro-squeezed transform (WSST). Knowl.-Based Syst. **208**, 106453 (2020)
19. Zhou, J., Yang, X., Zhang, L., Shao, S., Bian, G.: Multisignal VGG19 network with transposed convolution for rotating machinery fault diagnosis based on deep transfer learning. Shock. Vib. **2020**, 1–12 (2020)
20. Wang, Y., Ding, X., Zeng, Q., Wang, L., Shao, Y.: Intelligent rolling bearing fault diagnosis via vision ConvNet. IEEE Sens. J. **21**(5), 6600–6609 (2020)
21. Kaji, M., Parvizian, J., Venn, H.W.V.D.: Constructing a reliable health indicator for bearings using convolutional autoencoder and continuous wavelet transform. Appl. Sci. **10**, 8948 (2020)
22. Ma, P., Zhang, H., Fan, W., Wang, C., Wen, G., Zhang, X.: A novel bearing fault diagnosis method based on 2D image representation and transfer learning-convolutional neural network. Meas. Sci. Technol. **30**(5), 055402 (2019)
23. Li, X., Zhang, W., Ding, Q., Sun, J.Q.: Intelligent rotating machinery fault diagnosis based on deep learning using data augmentation. J. Intell. Manuf. **31**(2), 433–452 (2020)
24. Xu, G.W., Liu, M., Jiang, Z.F., Soffker, D., Shen, W.M.: Bearing fault diagnosis method based on deep convolutional neural network and random forest ensemble learning. Sensors **19**(5), 1424–8220 (2019)

25. Zhao, B., Zhang, X.M., Li, H., Yang, Z.B.: Intelligent fault diagnosis of rolling bearings based on normalized CNN considering data imbalance and variable working conditions. Knowl.-Based Syst. **199**, 105971 (2020)
26. Li, H., Zhang, Q., Qin, X., Sun, Y.: Raw vibration signal pattern recognition with automatic hyper-parameter-optimized convolutional neural network for bearing fault diagnosis. Proc. Inst. Mech. Eng. C J. Mech. Eng. Sci. **234**(1), 343–360 (2019)
27. Gu, J., Wang, Z., Kuen, J., Ma, L., Shahroudy, A., Shuai, B., Liu, T., Wang, X., Wang, G., Cai, J., Chen, T.: Recent advances in convolutional neural networks. Pattern Recogn. **77**, 354–377 (2018)
28. Aghdam, H.H., Heravi, E.J.: Guide to Convolutional Neural Networks. Springer, New York, NY 10(978–973), 51 (2017)
29. Lavin, A., Gray, S.: Fast algorithms for convolutional neural networks. In: Proceedings of the IEEE Conference on Computer Vision and Pattern Recognition, pp. 4013–4021 (2016)
30. Kiranyaz, S., Avci, O., Abdeljaber, O., Ince, T., Gabbouj, M., Inman, D.J.: 1D convolutional neural networks and applications: a survey. Mech. Syst. Signal Process. **151**, 107398 (2021)
31. Liu, M., Shi, J., Li, Z., Li, C., Zhu, J., Liu, S.: Towards better analysis of deep convolutional neural networks. IEEE Trans. Visual Comput. Graphics **23**(1), 91–100 (2016)
32. Khan, A., Sohail, A., Zahoora, U., Qureshi, A.S.: A survey of the recent architectures of deep convolutional neural networks. Artif. Intell. Rev. **53**(8), 5455–5516 (2020)
33. Liu, N., Han, J., Zhang, D., Wen, S., Liu, T.: Predicting eye fixations using convolutional neural networks. In: Proceedings of the IEEE Conference on Computer Vision and Pattern Recognition, pp. 362–370 (2015)
34. Bilal, A., Jourabloo, A., Ye, M., Liu, X., Ren, L.: Do convolutional neural networks learn class hierarchy? IEEE Trans. Visual Comput. Graph. **24**(1), 152–162 (2017)
35. Sharma, N., Jain, V., Mishra, A.: An analysis of convolutional neural networks for image classification. Proc. Comput. Sci. **132**, 377–384 (2018)
36. Jáureg, J.C., Reséndiz, J.R., Thenozhi, S., Szalay, T., Jacsó, Á., Takács, M.: Frequency and time-frequency analysis of cutting force and vibration signals for tool condition monitoring. IEEE Access **6**, 6400–6410 (2018)
37. Singh, B., Saboo, N., Kumar, P.: Effect of short-term aging on creep and recovery response of asphalt binders. J. Transp. Eng. Part B Pavements **143**(4), 04017017 (2017)
38. Chen, L., Zheng, L., Yang, J., Xia, D., Liu, W.: Short-term traffic flow prediction: from the perspective of traffic flow decomposition. Neurocomputing **413**, 444–456 (2020)
39. Rhif, M., Ben Abbes, A., Farah, I.R., Martínez, B., Sang, Y.: Wavelet transform application for/in non-stationary time-series analysis: a review. Appl. Sci. **9**(7), 1345 (2019)
40. Zhang, C.W., Yang, M.Y., Zeng, H.J., Wen, J.P.: Pedestrian detection based on improved LeNet-5 convolutional neural network. J. Algorithms Comput. Technol. **13**, 1748302619873601 (2019)
41. Smith, W.A., Randall, R.B.: Rolling element bearing diagnostics using the case western reserve university data: a benchmark study. Mech. Syst. Signal Process. **64**, 100–131 (2015)
42. Yazdi, M., Khan, F., Abbassi, R., Rusli, R.: Improved DEMATEL methodology for effective safety management decision-making. Saf. Sci. **127**, 104705 (2020). https://doi.org/10.1016/j.ssci.2020.104705
43. Adumene, S., et al.: Dynamic logistics disruption risk model for offshore supply vessel operations in Arctic waters. Marit. Transp. Res. **2**, 100039 (2021). https://doi.org/10.1016/j.martra.2021.100039
44. Golilarz, N.A., Gao, H., Pirasteh, S., Yazdi, M., Zhou, J., Fu, Y.: Satellite multispectral and hyperspectral image de-noising with enhanced adaptive generalized Gaussian distribution threshold in the wavelet domain. Remote Sens. **13**, 101 (2021). https://doi.org/10.3390/rs13010101
45. Golilarz, N.A., Mirmozaffari, M., Gashteroodkhani, T.A., Ali, L., Dolatsara, H.A., Boskabadi, A., Yazdi, M.: Optimized wavelet-based satellite image de-noising with multi-population differential evolution-assisted Harris hawks optimization algorithm. IEEE Access **8**, 133076–133085 (2020)

46. Kabir, S., Geok, T.K., Kumar, M., Yazdi, M., Hossain, F.: A method for temporal fault tree analysis using intuitionistic fuzzy set and expert elicitation. IEEE Access **8**, 980–996 (2020)
47. Kabir, S., Yazdi, M., Aizpurua, J.I., Papadopoulos, Y.: Uncertainty-aware dynamic reliability analysis framework for complex systems. IEEE Access **6**, 29499–29515 (2018)

Chapter 5
Using Global Average Pooling Convolutional Siamese Networks for Fault Diagnosis of Planetary Gearboxes

Abstract Planetary gearboxes have been maturely applied to a broad of scenarios in industry, and fault diagnosis is of great need for their design issues and operation and maintenance activities. This chapter introduces a Global Average Pooling-based Convolutional Siamese Network (GAPCSN) for the fault diagnosis of planetary gearboxes, which can cope with limited data situations. Initially, a convolutional layer with a wide convolutional kernel is used in the feature extraction module to improve the feature extraction capability of the model. Subsequently, a maximum pooling layer and a global average pooling layer are designed for the dimensionality reduction of extracted features and reducing the number of parameters of the network model. Euclidean distance is applied to quantify feature vectors to improve the classification capability of the model. The fault diagnoses of planetary gearboxes is carried out to validate the proposed model carry out the model validation, and the results show that GAPCSN performs better than other existing models in fault diagnosis of planetary gearboxes under limited data. Overall, the introduced fault diagnoses model, GAPCSN, contributes to condition-based maintenance and predictive maintenance of complex systems such as planetary gearboxes in energy infrastructure assets.

Keywords Fault diagnosis · Planetary gearboxes · GAPCSN · Condition-based maintenance

5.1 Introduction

Planetary gearboxes consist of multiple bearings, gears, and shafts, which are fundamental parts of energy infrastructure assets such as power plants and turbines [1–5]. Fault diagnosis for planetary gearboxes is more complicated than for a single bearing due to the complicity in structure [6, 7]. On the other hand, unlike the health state monitoring of bearings that can be conducted in laboratories, the health state monitoring of planetary gearboxes is difficult and costly. Accordingly, the sector lacks sufficient monitoring data on planetary gearboxes for fault diagnosis, condition-based maintenance, predictive maintenance, and health management.

The last decades have seen significant progress in fault diagnosis methodologies. Two leading solutions have been applied to the fault diagnosis with insufficient data [8–12]: (i) Migration learning and (ii) Model regularization and data enhancement techniques. Migration learning refers to acquiring knowledge, skills, and attitudes that enable individuals to migrate successfully to and integrate into a new cultural environment. This type of learning can be formal and informal and can occur before, during, and after the migration process. Before migration, individuals may engage in formal education, such as language classes, cultural orientation programs, and job training, to prepare for life in their destination country. Informal learning, such as learning from family members or friends who have previously migrated, can also help prepare for the migration. It is to migrate the model to a small number of samples for fine-tuning to obtain a fault diagnosis model with high recognition capability. Model regularization and data enhancement techniques are two standard methods used in machine learning to improve the performance of models. Model regularization refers to adding constraints to a model to prevent overfitting, which occurs when a model fits the training data too closely and performs poorly on new, unseen data. There are several methods of model regularization. Data enhancement techniques augment the available training data by creating recent training examples based on the existing data. This can increase the training set's size, reduce overfitting, and improve the model's generalization performance. There are several techniques for data enhancement. Those methods alleviate the problem of model overfitting under insufficient data conditions. The following lists some state-of-the-art studies and their applications.

Regarding migration learning methodologies, Chen et al. [13] proposed a migration learning-based fault diagnosis framework for missing data, which can fully use a large number of structurally incomplete samples to extract functional fault features through a suitable migration learning mechanism. Xiao et al. [14] proposed a migration learning-based fault diagnosis framework, which exhibited better fault diagnosis accuracy in insufficient target data. Wen et al. [15] proposed a TCNN-based migration learning model consisting of a deep network structure with a recognition accuracy of up to 99.99%. Moreover, Li et al. [16] proposed a data augmentation technique to artificially increase the effective samples used for model training, which exhibited high diagnostic accuracy on several original training samples. Dong et al. [17] proposed an intelligent diagnosis framework for rolling bearing bushing faults based on dynamic modeling and migration learning. The framework can reduce the feature distribution variance and improve the fault identification performance.

Regarding model regularization and data enhancement techniques, Luo et al. [18] proposed a fault diagnosis method based on conditional deep convolutional generative adversarial network generation models, which proved to be helpful in improving the model's recognition accuracy and generalization capability. Li et al. [19] proposed a fault diagnosis method for permanent magnet synchronous motors based on conditional generative adversarial networks. They optimized sparse autoencoders and improved the generalization and classification ability of the model while expanding the diversity of training samples. Although model regularization and data augmentation techniques improve the learning classification ability of the model to a certain extent, they fail to cope with the problem that the training samples carry little feature information, resulting in a weak feature extraction ability and low recognition accuracy of the final trained model.

For classification problems under insufficient data, twin neural networks are applied to solve classification problems [20]. Twin neural networks do not require extracting features from many samples, reducing the sample size. In recent years, twin neural networks based on metric learning have been applied to the fault diagnosis of mechanical equipment with insufficient data and intense noise. To be specific, Zhang et al. [21] proposed a bearing fault diagnosis algorithm based on small sample learning, which utilizes twin neural networks for metric learning from input samples and achieves high diagnostic accuracy under limited samples, noises, and variable operating conditions. Tuyet-Doan et al. [22] proposed a one-time learning method. They applied it to fault diagnosis of gas-insulated switchgear with limited training samples, which overcame the problem of low recognition accuracy of deep learning algorithms caused by insufficient data.

The above study shows that twin neural networks perform better fault diagnosis under limited training samples and noise conditions but have not yet been applied to fault diagnosis of planetary gearboxes. Therefore, this chapter designs a convolutional Siamese network based on global average pooling for fault diagnosis of planetary gearboxes under insufficient data. The model uses a convolutional layer with a wide convolutional kernel in the first layer to extract features from the input data. The extracted features are then imported to the global average pooling layer to obtain feature vectors and reduce the model's parameters to improve the model's generalization ability. The Euclidean distance is used to learn the metric of the feature vector classification.

The rest of the chapter is arranged as follows. Section 5.2 presents the fault diagnosis method. A case is presented in Sect. 5.3. The conclusions are provided in Sect. 5.4.

5.2 Methods

5.2.1 Convolutional Siamese Networks Based on Global Average Pooling

This chapter designs Convolutional Siamese Networks with Global Average Pooling (GAPCSN) for fault diagnosis of planetary gearboxes under insufficient data conditions. The network uses the convolutional neural network for the feature extraction, and Euclidean distance is applied for the similarity metric.

Convolutional neural networks mainly consist of feature extraction and classification modules [23–25]. The feature extraction module extracts the features from the input data and consists of a convolutional layer, a pooling layer, and an activation layer [23]. The classification module identifies and classifies the extracted features, which consist of fully connected layers.

The convolutional neural network's convolutional layer uses weight sharing for convolutional computation. Its mathematical model is:

$$y^{l(i,j)} = K_i^l * x^{l(r^j)} = \sum_{j'=0}^{W-1} K_i^{l(j')} x^{l(j+j')} \tag{5.1}$$

where, $K_i^{l(j')}$ denotes the j' weight of the convolution kernel i in the layer l, $*$ denotes the convolution operation, $x^{l(r^j)}$ represents the convolved local region j in the layer l, W is the width of the convolution kernel, $y^{l(i,j)}$ and reflects the output.

The activation layer uses an activation function to perform a non-linear transformation on logit's values of convolution output. It enhances the multidimensional features' linear separability [25]. The Sigmoid function, Rectified Linear Unit (ReLU), is always used as an activation function. The mathematical models of the activation functions are:

$$a^{l(i,j)} = Sigmoid\left(y^{l(i,j)}\right) = \frac{1}{1+e^{-y^{l(i,j)}}} \tag{5.2}$$

$$a^{l(i,j)} = f\left(y^{l(i,j)}\right) = \max\left\{0,\, y^{l(i,j)}\right\} \tag{5.3}$$

where $a^{l(i,j)}$ denotes the activation value of the convolutional layer output $y^{l(i,j)}$.

The pooling layer is also known as the sampling layer [24]. The maximum pooling method is used as the down-sampling algorithm in the pooling layer, which uses a filtering window to take the maximum value of the input elements as the output of the maximum pooling by:

$$y_i^{l+1}(j) = \max_{k \in D_j}\left\{x_i^l(k)\right\} \tag{5.4}$$

where $x_i^l(k)$ denotes the element of the feature i in the layer l in the range of the pooling kernel, $y_i^{l+1}(j)$ denotes the element in the feature map i of the layer $l+1$ after pooling, D_j and denotes the pooling region j.

Global averaging pooling averages feature data (after convolution) and exports them to the next layer [20] as:

$$y_k = \frac{1}{|R|} \sum_{(p,q) \in R} x_{kpq} \tag{5.5}$$

where y_k denotes the global average pooled output of the feature map kx_{kpq}, the element located (p,q) in the R feature map k region, and $|R|$ the number of elements in the feature map k.

The primary function of the fully-connected layer is to identify and classify the extracted features [25–28]. The mathematical model of the forward propagation of the fully-connected layer is as follows:

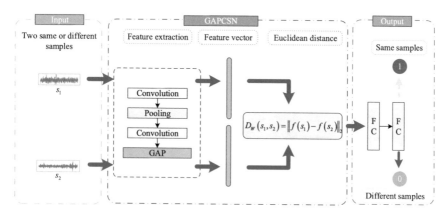

Fig. 5.1 The structure of the fault diagnosis model

$$z^{l+1(j)} = \sum_{i=1}^{n} W_{ij}^l a^{l(i)} + b_j^l \tag{5.6}$$

where W_{ij}^l is the weight between the neuron i in layer l and neuron j in the layer $l+1$, z_j^{l+1} is the logit value of the output neuron j in the layer $l + 1 b_j^l$, and denotes the bias of neurons in layer l to the j neuron in the layer $l + 1$.

The similarity measurement achieves classification by measuring the similarity between unknown and known labeled samples. The distance between two sample feature vectors is calculated using Euclidean distance [29] to rate the similarity between two samples. The Euclidean distance between two sample eigenvectors $f(s_1)$ $f(s_2)$ is:

$$D_W(s_1, s_2) = \|f(s_1) - f(s_2)\|_2 = \left(\sum_{i=1}^{p} (f(s_1^i) - f(s_2^i))^2 \right)^{\frac{1}{2}} \tag{5.7}$$

where, s_1 and s_2 denote input samples, $D_W(s_1, s_2)$ is the Euclidean distance between the input sample feature vectors, $f(s_1)$ and $f(s_2)$ denote the feature vectors of the two input samples, P is the feature dimension of the samples and $(f(s_1^i) - f(s_2^i))$ is the coordinate difference between $f(s_1)$ and $f(s_2)$ on the feature dimension i.

The GAPCSN-based fault diagnosis model is shown in Fig. 5.1.

5.2.2 GAPCSN-Based Fault Diagnosis

The fault diagnosis process based on GAPCSN, see Fig. 5.2, consists of three steps: data pre-processing, model training, model testing, and diagnosis.

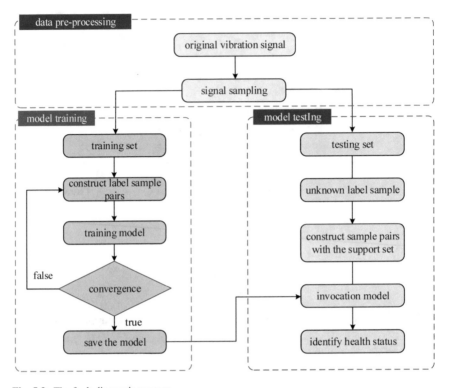

Fig. 5.2 The fault diagnosis process

Step 1: Data pre-processing

Step 1.1: Collect planetary gearbox vibration signals through the sensor.

Step 1.2: Sampling the vibration signals using the sliding window to get training samples and test samples.

Step 1.3: Normalize the collected signal x by:

$$\tilde{x} = \frac{x - x_{\min}}{x_{\max} - x_{\min}} \tag{5.8}$$

where, x is the data before normalization, $\tilde{x}$ is the data after normalization; $x_{\min}$ and $x_{\max}$ are the minimum and maximum values.

Step 2: Model training

In the training phase of GAPCSN, the input data is generated by random matching from the training set. The input data is forward propagated to compute a prediction value. The loss function calculates the error between the predicted and actual values.

The convolutional Siamese network updates the network weights using the backprop-agation algorithm to reduce the error between the predicted and actual values. The training completes when the model converges. The trained network model is used to identify unknown samples type. The output of the model represents the similarity of two input samples, and its mathematical model is:

$$P(s_1, s_2) = Sigm(FC(D_W(s_1, s_2))) \tag{5.9}$$

where, s_1 and s_2 denote input samples, $D_W(s_1, s_2)$ is the Euclidean distance between input sample feature vectors, $Sigm$ is the activation function, FC is the fully connected layer, $P(s_1, s_2)$ is the similarity of the input samples s_1 and s_2.

The loss function of the model is a regularized cross-entropy, represented by:

$$L(s_1, s_2, y) = y \log(P(s_1, s_2)) + (1 - y) \log(1 - P(s_1, s_2)) \tag{5.10}$$

where, s_1 and s_2 denote input samples; y is the label values corresponding to the input samples, $P(s_1, s_2)$ denotes the similarity of the input samples s_1 and s_2.

The model's parameters are optimized using the Adam optimization algorithm that holds robustness for the choice of hyperparameters:

$$m_w^{(T+1)} = \beta_1 m_w^{(T)} + (1 - \beta_1)\nabla_w L^{(T)} \tag{5.11}$$

$$v_w^{(T+1)} = \beta_2 v_w^{(T)} + (1 - \beta_2)(\nabla_w L^{(T)})^2 \tag{5.12}$$

$$m = \frac{m_w^{(T+1)}}{1 - (\beta_1)^{T+1}} \tag{5.13}$$

$$v = \frac{v_w^{(T+1)}}{1 - (\beta_2)^{T+1}} \tag{5.14}$$

$$w^{(T+1)} = w^{(T)} - \eta \frac{\hat{m}_w}{\sqrt{\hat{v}_w} + \varepsilon} \tag{5.15}$$

where, $w^{(T)}$ is the generation T parameter, $L^{(T)}$ is the loss function, β_1 is the forgetting factor of the first order moment of the gradient, β_2 denotes the forgetting factor of the second order moment of the gradient m, and v is the moving average.

Step 3: Model testing and diagnosis

The testing of GAPCSN is carried out by metric learning of the test and support datasets, where the test set consists of samples of unknown types, and the support set contains various labeled samples. The testing process of the convolutional Siamese network is known as the K-shot-M-way strategy, which is implemented by metric learning of test and support sets.

In the testing phase of GAPCSN, the input data is a sample pair consisting of test samples in the test set and labeled samples in the support set. The input data are extracted by feature extraction, and the extracted feature vectors are subjected to a distance metric. They are classified as the same type if the metric distance is less than a threshold. The test samples are identified by counting the similarity between the test samples and each labeled sample class. The test samples are categorized into labeled samples with the highest similarity. This chapter performs the K-group One-shot M-way test to simulate the implementation and calculates the maximum probability sum of the same label as:

$$C(\hat{s}, (T)) = \arg\max_{c}\left(\sum_{n=1}^{K} p(\hat{s}, s_{cn})\right), s_{cn} \in T_n \tag{5.18}$$

where, $\hat{s}$ is $\hat{s}$ unknown failure samples T_n denote the support set, s_{cn} is the sample in the support set T_n that is similar to $\hat{s}$.

5.3 Illustrative Cases and Results

5.3.1 Data Description

This chapter focuses on the fault diagnosis of planetary gearboxes, and an open dataset [30] is used to verify the model's performance. The planetary gearbox dataset contains the monitoring data of the planetary gearbox working at a speed of 1200 rpm and a load of 0 Nm. This includes five health conditions of the sun wheel: normal, missing teeth, broken teeth, cracked tooth roots, and worn tooth surfaces. The monitoring data is a vibration signal collected by a three-way acceleration sensor with a sampling frequency of 5.12 kHz.

This chapter uses 1024 sampling points as a sliding window to sample the monitoring data of five health states in X direction without overlap. In this chapter, 85 samples are collected for each health state of the solar wheel. They are divided into 60 training samples and 25 test samples according to the ratio of 7:3. Therefore, the five health states of the solar wheel contain a total of 300 training samples and 125 test samples. "normal", "missing tooth", "broken tooth," "tooth root crack," and "tooth surface wear" are labeled as 0, 1, 2, 3 and 4, respectively. The sampled dataset is dataset A, as shown in Table 5.1. Similarly, the monitoring data in Y and Z directions are sampled without overlap as datasets B and C.

Table 5.1 Planetary gearbox failure data

Dataset	Types	Label	Description	Direction	Training samples	Testing samples
Dataset A	Health	0	Healthy operating condition	X	60	25
	Chipped	1	A crack occurs in the feet		60	25
	Miss	2	One of the feet is missing		60	25
	Root	3	A crack occurs in the root of the feet		60	25
	Surface	4	The wear occurs on the surface		60	25
Dataset B	Health	0	Healthy operating condition	Y	60	25
	Chipped	1	A crack occurs in the feet		60	25
	Miss	2	One of the feet is missing		60	25
	Root	3	A crack occurs in the root of the feet		60	25
	Surface	4	The wear occurs on the surface		60	25
Dataset C	Health	0	Healthy operating condition	Z	60	25
	Chipped	1	A crack occurs in the feet		60	25
	Miss	2	One of the feet is missing		60	25
	Root	3	A crack occurs in the root of the feet		60	25
	Surface	4	The wear occurs on the surface		60	25

5.3.2 Network Parameter Design

The global average pooling-based convolutional Siamese network consists of feature extraction and similarity metric, see Table 5.2. To be specific, the feature extraction part of GAPCSN adopts a convolutional neural network structure composed of two layers of convolution and pooling, whose parameters are designed as follows:

Table 5.2 The specific structure of GAPCSN

No.	Layer type	Kernel size	Kernel number	Kernel number	Padding
1	Conv1	64×1	16	16	Same
2	MaxPool1	2×1	2	16	Valid
3	Conv2	3×1	1	32	Same
4	GAP	–	–	–	Valid
5	Lambda	–	–	–	–
6	FC1	–	–	100	–
7	FC2	–	–	1	–

(i)   The first convolutional layer adopts 16 wide convolutional kernels with the size of 64×1 to perform convolutional operations on the time-domain signal; The convolutional layer uses 16 steps as the convolutional step; during the convolution, the zero-complement operation is performed on the edges of the input matrix; after the convolutional operation, the ReLU function is used to perform non-linear activation on the convolutional output.

(ii) The first pooling layer of the convolutional neural network uses maximum pooling as the down-sampling operation. This pooling layer consists of 16 pooling kernels with the size of 2×1.

(iii) The second convolutional layer of the convolutional neural network consists of 32 convolutional kernels with the size of 3×1. This convolutional kernel performs a convolutional operation at every step. A zero-complement operation is performed on the edges of the input matrix. After the convolutional operation, the ReLU function performs non-linear activation on the convolutional output.

(iv) The second pooling layer uses global average pooling as the down-sampling operation. The global average pooling layer replaces the fully connected layer to reduce the number of model training parameters, decrease the computational effort of the model, reduce the probability of overfitting the model, and convert the convolutional output into feature vectors.

The similarity metric of GAPCSN uses Euclidean distance and two fully-connected layers to calculate the similarity between two input samples with the following parameters:

(i)   The first layer of the similarity metric uses Euclidean distance to measure the distance between the feature vectors extracted from the two samples.

(ii) The second layer of the similarity metric uses a fully connected layer with 100 neurons to convert the results into vectors.

(iii) The final fully connected layer uses one neuron to calculate the probability value of the similarity of the two input samples.

5.3.3 Model Training

To verify the fault diagnosis performance of the GAPCSN model under insufficient data conditions and evaluate the effect of sample size on the model performance, this chapter randomly selects 30, 60, 120, 180, 240, and 300 training samples from the training samples of dataset A as training sets for comparison tests. To avoid the bias caused by randomly selected training samples, this chapter performs five random selections of training samples for each training set. It uses each randomly selected training sample to train the GAPCSN model. During the training process, the Adam optimization algorithm is used to update the parameters of the GAPCSN model with a learning rate of 0.001.

To further validate the advances of the GAPCSN model, Support Vector Machine (SVM), Deep Convolutional Neural Network (WDCNN), and Small Sample Learning method (FSL) are trained under the training set for comparison reasons.

To further evaluate the robustness and generalization ability of the GAPCSN model in monitoring data from different directions for the same working condition, the training samples of dataset B and dataset C are used to train the GAPCSN and other models. Therefore, three datasets are used to validate the GAPCSN model.

5.3.4 Model Testing

125 test samples from dataset A from this chapter are regarded as the test set. The testing process adopts the Five-shot, Five-way strategy, which implements the test set with a support set consisting of five fault types and five labeled samples of each class for metric learning. To avoid training bias affecting the performance evaluation of the model, the model is tested after each training, and the test average is taken as the final recognition result. Similarly, the test samples in datasets B and C are test sets. The models trained in both datasets are tested separately according to the same test as in dataset A.

5.3.5 Results

5.3.5.1 Test Results Based on Dataset A

The test accuracies of the four algorithms based on dataset A are shown in Fig. 5.3, in which the horizontal coordinate indicates the training sets consisting of different numbers of training samples, and the vertical coordinate is the test accuracy of the model. It can be seen that under the minimal sample condition with 30 training samples, the test accuracy of GAPCSN is 98.40% and outperforms 47.92% of WDCNN, 72.56% of FSL, and 85.76% of SVM. It indicates that GAPCSN shows

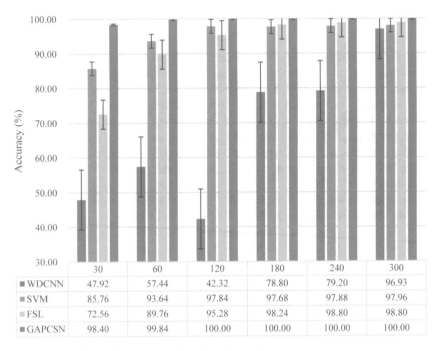

	30	60	120	180	240	300
■ WDCNN	47.92	57.44	42.32	78.80	79.20	96.93
■ SVM	85.76	93.64	97.84	97.68	97.88	97.96
■ FSL	72.56	89.76	95.28	98.24	98.80	98.80
■ GAPCSN	98.40	99.84	100.00	100.00	100.00	100.00

Fig. 5.3 Comparison of test accuracies based on dataset A

better recognition performance than other algorithms under limited training samples. With the increase of training samples, the test accuracy of each algorithm is steadily improved, but the GAPCSN maintains a higher accuracy than other algorithms. It is noted that the test accuracy of GAPCSN reaches 100% when the number of training samples increases to 120.

5.3.5.2   Test Results Based on Dataset B

The test accuracies of the four algorithms based on dataset B are shown in Fig. 5.4. Under the minor sample condition with 30 training samples, GAPCSN achieved a test accuracy of 96.88%, which is higher than 42.64% of WDCNN, 37.68% of FSL, and 66.96% of SVM. With the increase of training samples, the test accuracy of each algorithm is gradually improved, and the recognition accuracy of GAPCSN is higher than that of other algorithms under all scenarios. It is highlighted that when the number of training samples increases to 300, the test accuracy of each algorithm is the highest. For instance, the test accuracy of GAPCSN is 98.27%. At the same time, when the training set consists of 30, 60, 120, and 180 samples, the test accuracy of all algorithms except GAPCSN is lower than 90%, which can be inferred that the data quality of dataset B is lower than that of dataset A. There is a specific interference signal in the data monitored by the sensor in the Y direction, which has

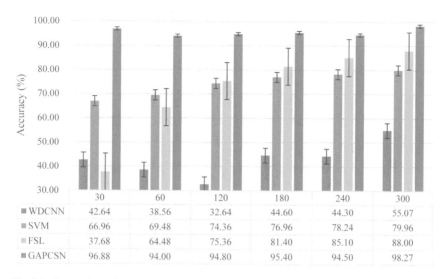

Fig. 5.4 Comparison of test accuracies based on dataset B

a particular impact on the recognition performance of algorithms. It shows that the data quality seriously affects the model's recognition performance. Despite the poor quality of the Y-direction monitoring data, GAPCSN can recognize faults effectively, and the recognition accuracy of which is consistently above 94%, which verifies that GAPCSN holds better robustness and generalization ability than other algorithms.

5.3.5.3 Test Results Based on Dataset C

The test accuracies of the four algorithms based on dataset C are shown in Fig. 5.5. Under the minor sample condition with 30 samples, the test accuracy of GAPCSN is 96.64%, which is better than 49.12% of WDCNN, 65.6% of FSL, and 84.40% of SVM. As the number of training samples increased, the test accuracy of each algorithm improved. When the number of training samples increases to 60, the test accuracy of GAPCSN is 100%.

The samples in datasets A, B, and C are obtained from the accelerometer monitoring samples from three directions: X, Y, and Z. To further analyze the robustness and generalization ability of GAPCSN, the test results of GAPCSN are synthesized. The test accuracy of GAPCSN in the three directions monitoring data is shown in Fig. 5.6. With the increase in the number of training samples, the test accuracy of GAPCSN shows a steady improvement in general but fluctuates up and down in a small range. For example, in the Y-direction's monitoring data, the test accuracy of GAPCSN was first dropped from 96.88 to 94.00%, then gradually increased to 98.27%. When the number of training samples increased to 300, the test accuracy of GAPCSN reached the highest level, 98%. In the extreme case with 30 training samples, the test accuracy of GAPCSN in the three directions' monitoring

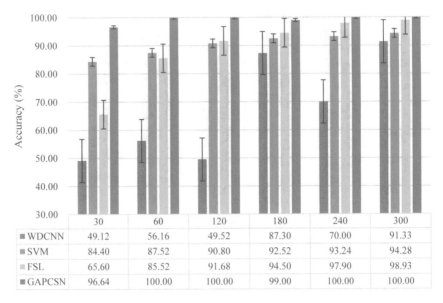

	30	60	120	180	240	300
▪ WDCNN	49.12	56.16	49.52	87.30	70.00	91.33
▪ SVM	84.40	87.52	90.80	92.52	93.24	94.28
▪ FSL	65.60	85.52	91.68	94.50	97.90	98.93
▪ GAPCSN	96.64	100.00	100.00	99.00	100.00	100.00

Fig. 5.5   Comparison of test accuracies based on dataset C

data is 98.40, 96.88, and 96.64%. In summary, it can be illustrated that GAPCSN can show high diagnostic ability for monitoring data in all three directions under limited training samples. Its recognition accuracy does not fluctuate drastically due to the change in the number of training samples, showing better robustness and generalization ability.

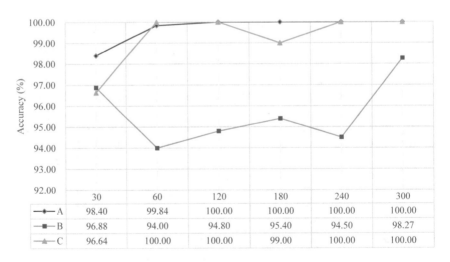

	30	60	120	180	240	300
◆ A	98.40	99.84	100.00	100.00	100.00	100.00
▪ B	96.88	94.00	94.80	95.40	94.50	98.27
▲ C	96.64	100.00	100.00	99.00	100.00	100.00

Fig. 5.6   Test accuracy in three directions

5.3.5.4   Results Visualization

To evaluate the performance of GAPCSN more intuitively, this chapter visualizes the original monitoring data and the test results of GAPCSN in three directions, using the training set consisting of 60 training samples in datasets A, B, and C. To facilitate the representations: "0" refers to the normal state of the sun gear, "1" refers to the defect fault of the sun gear, "2" refers to the broken tooth fault of the sun gear, "3" refers to the tooth root wear fault of the sun gear, and "4" refers to the tooth surface wear failure of the generation sun gear.

To understand the distribution of monitoring data in three directions, this chapter uses t-SNE to downscale and visualize the original data of the three datasets, see Fig. 5.7. The horizontal and vertical coordinates of each figure indicate the characteristic values of the original data in the two-dimensional space, respectively.

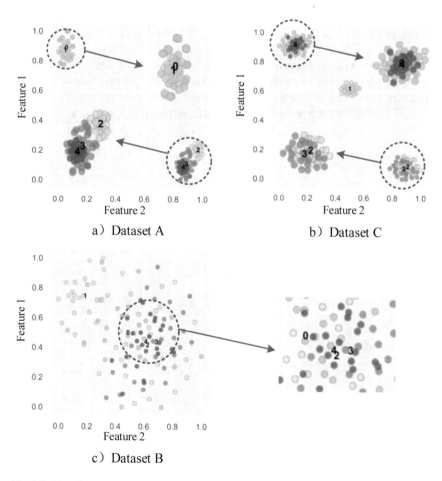

Fig. 5.7 Visualization of original data

As shown in Fig. 7a, the sample data in the X-direction shows more obvious feature information after dimensionality reduction. The distribution is more concentrated, among which the feature data with labels 0 and 1, 3 and 4 are mixed tog cannot be separated. In contrast, the feature data with labels 0 and 2 are better separated.

As shown in Fig. 7b, the sample in Z-direction also shows more feature information after dimensionality reduction. The distribution is more concentrated, in which the feature data with labels 0 and 4, 2 and 3 are mixed, while the feature data with labels 0 and 2 are better separated.

In Fig. 7c, the sample data in the Y-direction shows disordered feature information after dimensionality reduction. The feature data with each label cannot be clustered and separated.

From the visualization point of view, datasets A and C have better data separability and are more suitable for fault diagnosis than dataset B. It shows that the data sampled by the accelerometer monitoring from X and Z directions are of higher quality than that in the Y direction. Similarly, the analysis of features is displayed in Fig. 5.8. The analysis of Fig. 5.8 is the same as that of Fig. 5.7. Hence, it is not extended in this chapter.

As shown in Fig. 9a, based on dataset A, the feature data extracted by GAPCSN has better classification effects on the two-dimensional space. Based on dataset B, Fig. 9b indicates that the original data with labels 2 and 3 are overwhelmingly separated on the two-dimensional space by the feature data extracted by GAPCSN. Figure 9c,

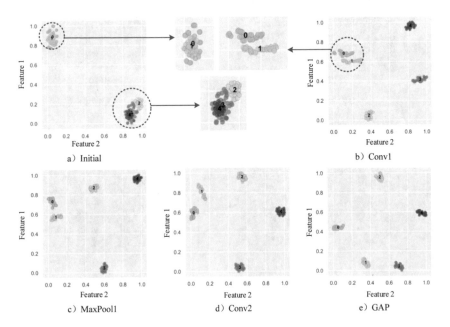

Fig. 5.8 Visualization of features

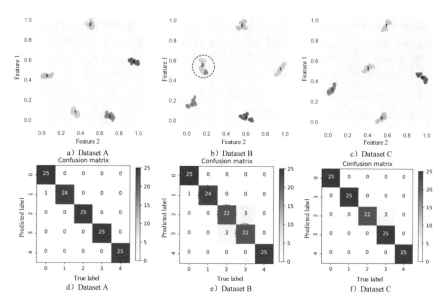

Fig. 5.9 Visualization of test results

based on dataset C, indicates that the feature data extracted by GAPCSN achieves better classification in two-dimensional space.

There are 25 test samples under each class of labels, so the central diagonal values of the confusion matrix indicate the number of test samples correctly recognized by the model in each category. As shown in Fig. 9d–f, GAPCSN holds good diagnostic performance in all three datasets. For instance, in dataset B with poor data quality, for the 25 test samples with label 1, GAPCSN recognize 24 successfully. Similarly, it recognizes 22 from the 25 samples with labels 2 and 3.

5.4 Conclusion

This chapter introduces a Global Average Pooling-based Convolutional Siamese Network (GAPCSN) for the fault diagnosis of planetary gearboxes, which can cope with limited data situations. A convolutional layer with a wide convolutional kernel is used in the feature extraction module to improve the feature extraction capability of the model. A maximum pooling layer and a global average pooling layer are designed for the dimensionality reduction of extracted features and reducing the number of parameters of the network model. Euclidean distance is applied to quantify feature vectors to improve the classification capability of the model. The fault diagnoses of planetary gearboxes carry out the model validation, and the results show that GAPCSN performs better in fault diagnosis of planetary gearboxes under limited data. Overall, the introduced fault diagnoses model, GAPCSN, contributes to condition-based maintenance and predictive maintenance of complex systems such as planetary gearboxes in energy infrastructure assets. However, as a direction for future study, it can be integrated with other types of intelligent tools into the energy infrastructure, which some have done by authors previously [31–37].

References

1. Lei, Y., Lin, J., Zuo, M.J., He, Z.: Condition monitoring and fault diagnosis of planetary gearboxes: a review. Measurement **48**, 292–305 (2014)
2. Li, H., Peng, W., Huang, C.G., Guedes Soares, C.: Failure rate assessment for onshore and floating offshore wind turbines. J. Marine Sci. Eng. **10**(12), 1965 (2022)
3. Li, H., Teixeira, A.P., Guedes Soares, C.: An improved failure mode and effect analysis of floating offshore wind turbines. J. Marine Sci. Eng. **10**(11), 1616 (2022)
4. Li, H., Yazdi, M.: Developing failure modes and effect analysis on offshore wind turbines using two-stage optimization probabilistic linguistic preference relations. In: Advanced Decision-Making Methods and Applications in System Safety and Reliability Problems. Studies in Systems, Decision and Control, vol. 211. Springer, Cham (2022). https://doi.org/10.1007/978-3-031-07430-1_4
5. Li, H., Huang, C.G., Soares, C.G.: A real-time inspection and opportunistic maintenance strategies for floating offshore wind turbines. Ocean Eng. **256**, 111433 (2022)
6. Li, H., Yazdi, M.: Reliability analysis of correlated failure modes by transforming fault tree model to Bayesian network: a case study of the MDS of a CNC machine tool. in: advanced decision-making methods and applications in system safety and reliability problems. In: Studies in Systems, Decision and Control, vol. 211. Springer, Cham (2022)
7. Lei, Y., Kong, D., Lin, J., Zuo, M.J.: Fault detection of planetary gearboxes using new diagnostic parameters. Meas. Sci. Technol. **23**(5), 055605 (2012)
8. Yazdi, M., Mohammadpour, J., Li, H., Huang, H.-Z., Zarei, E., Pirbalouti, R.G., Adumene, S.: Fault tree analysis improvements: a bibliometric analysis and literature review. Qual. Reliab. Eng. Int. n/a (2023). https://doi.org/10.1002/qre.3271
9. Li, H., Yazdi, M.: Advanced Decision-Making Methods and Applications in System Safety and Reliability Problems, Springer, Cham 2022. https://link.springer.com/book/9783031074295.
10. Li, H., Díaz, H., Soares, C.G.: A failure analysis of floating offshore wind turbines using AHP-FMEA methodology. Ocean Eng. **234**, 109261 (2021)
11. Li, H., Soares, C.G., Huang, H.Z.: Reliability analysis of a floating offshore wind turbine using Bayesian networks. Ocean Eng. **217**, 107827 (2020)
12. Wang, C., Li, H., Zhang, K., Hu, S., Sun, B.: Intelligent fault diagnosis of planetary gearbox based on adaptive normalized CNN under complex variable working conditions and data imbalance. Measurement **180**, 109565 (2021)
13. Chen, D., Yang, S., Zhou, F.: Transfer learning-based fault diagnosis with missing data due to multi-rate sampling. Sensors **19**(8), 1826 (2019)
14. Xiao, D., Huang, Y., Qin, C., Liu, Z., Li, Y., Liu, C.: Transfer learning with convolutional neural networks for small sample size problem in machinery fault diagnosis. Proc. Inst. Mech. Eng. C J. Mech. Eng. Sci. **233**(14), 5131–5143 (2019)
15. Wen, L., Li, X., Gao, L.: A transfer convolutional neural network for fault diagnosis based on ResNet-50. Neural Comput. Appl. **32**, 6111–6124 (2020)
16. Li, X., Zhang, W., Ding, Q., Sun, J.Q.: Intelligent rotating machinery fault diagnosis based on deep learning using data augmentation. J. Intell. Manuf. **31**, 433–452 (2020)
17. Dong, Y., Li, Y., Zheng, H., Wang, R., Xu, M.: A new dynamic model and transfer learning based intelligent fault diagnosis framework for rolling element bearings race faults: solving the small sample problem. ISA Trans. **121**, 327–348 (2022)

18. Luo, J., Huang, J., Li, H.: A case study of conditional deep convolutional generative adversarial networks in machine fault diagnosis. J. Intell. Manuf. **32**, 407–425 (2021)
19. Li, Y., Wang, Y., Zhang, Y., Zhang, J.: Diagnosis of inter-turn short circuit of permanent magnet synchronous motor based on deep learning and small fault samples. Neurocomputing **442**, 348–358 (2021)
20. Chopra, S., Hadsell, R., LeCun, Y.: Learning a similarity metric discriminatively, with application to face verification. In: 2005 IEEE Computer Society Conference on Computer Vision and Pattern Recognition (CVPR'05), vol. 1, pp. 539–546. IEEE (2005)s
21. Zhang, A., Li, S., Cui, Y., Yang, W., Dong, R., Hu, J.: Limited data rolling bearing fault diagnosis with few-shot learning. IEEE Access **7**, 110895–110904 (2019)
22. Tuyet-Doan, V.N., Do, T.D., Tran-Thi, N.D., Youn, Y.W., Kim, Y.H.: One-shot learning for partial discharge diagnosis using ultra-high-frequency sensor in gas-insulated switchgear. Sensors **20**(19), 5562 (2020)
23. Gu, J., Wang, Z., Kuen, J., Ma, L., Shahroudy, A., Shuai, B., Liu, T., Wang, X., Wang, G., Cai, J., Chen, T.: Recent advances in convolutional neural networks. Pattern Recogn. **77**, 354–377 (2018)
24. Aghdam, H.H., Heravi, E.J. Guide to Convolutional Neural Networks, vol. 10, no. 978–973, p. 51. Springer, New York, NY (2017)
25. Kuo, C.C.J.: Understanding convolutional neural networks with a mathematical model. J. Vis. Commun. Image Represent. **41**, 406–413 (2016)
26. Ghosh, A., Sufian, A., Sultana, F., Chakrabarti, A., De, D.: Fundamental concepts of convolutional neural network. In: Recent Trends and Advances in Artificial Intelligence and Internet of Things, pp. 519–567 (2020)
27. Liu, M., Shi, J., Li, Z., Li, C., Zhu, J., Liu, S.: Towards better analysis of deep convolutional neural networks. IEEE Trans. Visual Comput. Graphics **23**(1), 91–100 (2016)
28. Zhou, D.X.: Theory of deep convolutional neural networks: downsampling. Neural Netw. **124**, 319–327 (2020)
29. Hidayat, R., Yanto, I.T.R., Ramli, A.A., Fudzee, M.F.M., Ahmar, A.S.: Generalized normalized Euclidean distance based fuzzy soft set similarity for data classification. Comput. Syst. Sci. Eng. **38**(1), 119–130 (2021)
30. Shao, S., McAleer, S., Yan, R., Baldi, P.: Highly accurate machine fault diagnosis using deep transfer learning. IEEE Trans. Ind. Inf. **15**(4), 2446–2455 (2018)
31. Yazdi, M., Khan, F., Abbassi, R., Rusli, R.: Improved DEMATEL methodology for effective safety management decision-making. Saf. Sci. **127**, 104705 (2020)
32. Adumene, S., et al.: Dynamic logistics disruption risk model for offshore supply vessel operations in Arctic waters. Marit. Transp. Res. **2**(November), 100039 (2021)
33. Golilarz, N.A., Gao, H., Pirasteh, S., Yazdi, M., Zhou, J., Fu, Y.: Satellite multispectral and hyperspectral image de-noising with enhanced adaptive generalized Gaussian distribution threshold in the wavelet domain. Remote Sens. **13**, 101 (2021). https://doi.org/10.3390/rs1 3010101
34. Golilarz, N.A., Mirmozaffari, M., Gashteroodkhani, T.A., Ali, L., Dolatsara, H.A., Boskabadi, A., Yazdi, M.: Optimized wavelet-based satellite image de-noising with multi-population differential evolution-assisted Harris Hawks optimization algorithm. IEEE Access **8**, 133076–133085 (2020)
35. Kabir, S., Geok, T.K., Kumar, M., Yazdi, M., Hossain, F.: A method for temporal fault tree analysis using intuitionistic fuzzy set and expert elicitation. IEEE Access **8**, 980–996 (2020)
36. Kabir, S., Yazdi, M., Aizpurua, J.I., Papadopoulos, Y.: Uncertainty-aware dynamic reliability analysis framework for complex systems. IEEE Access **6**, 29499–29515 (2018)
37. Pirbalouti, R.G., Dehkordi, M.K., Mohammadpour, J., Zarei, E., Yazdi, M.: An advanced framework for leakage risk assessment of hydrogen refueling stations using interval-valued spherical fuzzy sets (IV-SFS). Int. J. Hydrogen Energy (2023). https://doi.org/10.1016/j.ijh ydene.2023.03.028.

Chapter 6
Advances in Failure Prediction of Subsea Components Considering Complex Dependencies

Abstract The technological advancement in subsea system design for harsh environments presents structural and functional performance complexity. This paper explores the various methodologies for failure assessment of subsea systems considering the unstable operating environment and functional dependencies among subcomponents. The various failure influencing factors for harsh offshore operations were examined to establish their level of importance and impact on the failure model prediction for the subsea assets. The systematic study explores the failure methods that integrate the data-driven approaches with the physics of failure models to enhance better failure risk prediction in a dynamic offshore environment. While presenting the state of knowledge and the advances in the failure prediction approaches, the study identified the need to further integrate the data-driven machine learning model with the multi-hazard failure risk aggregation approaches for a holistic safety criteria formulation for harsh subsea operations. This could be further supported by integrating resilient design, digitalization, and IoT for remote arctic subsea operations.

Keywords Failure prediction · Subsea systems · Functional dependencies · Harsh environments · Data-driven machine learning · Safety

6.1 Introduction

Subsea systems are critical infrastructures for sustainable offshore oil and gas operations. The systems' functionality and performance are critical to the safety of oil and gas operations, especially in harsh environments. There is no doubt that subsea systems are exposed to harsh environments. Besides the complexity of the subsea system technology, the harshness of the operating environment plays a role in the system's reliability over time [1]. The abiotic and biotic characteristics of the offshore environment instigate complex interactions that affect the structural integrity of the subsea infrastructure. Despite the technological advancement in subsea system design, there have been recent failures during operations [2–10].

H. Li et al., *Intelligent Reliability and Maintainability of Energy Infrastructure Assets*,
Studies in Systems, Decision and Control 473,
https://doi.org/10.1007/978-3-031-29962-9_6

Understanding the complexities in the system design and environmental parameters instability is fundamental to reliable failure prediction of the subsea infrastructure. Researchers have explored different qualitative, quantitative, and dynamic studies to better understand subsea systems' structural complexity and failure mechanisms. These studies categorized organizational, engineering, environmental, and operational factors as key failure-instigating factors of the subsea system. For instance, poor integrity management structure, corrosion, fatigue, and human error have been identified as instigating failure mechanisms in offshore oil and gas operations. The basic actions and environmental factors that result in the subsea systems' degradation can be linked with these identified failure mechanisms.

Furthermore, the failure-instigating factors could exhibit different correlations and dependencies. Complex dependencies among key system parameters impact the overall performance of such systems. A proper understanding of the structural interactions among the functional parameters is crucial for a reliable model structural development and performance prediction. Existing advanced failure assessment techniques, such as failure mode and effect analysis (FMEA), have been used to define the likely deviation of subsea oil and gas systems. The essence is to technically explore the subsystem performance considering the system's complexity and criticality, such as the blowout preventer (BOP) [11]. To further explore the inductive structural failure of the subsea system, Yuhua and Datao [12] adopt a fuzzy-based fault tree structure. The authors identified the various failure modes of subsea pipelines and developed a failure propagation using fault tree analysis (FTA). Fifty-five minimal cut sets were identified, and their probabilities were estimated using fuzzy-based elicitation. Cheliyan and Bhattacharyya [13] explored the benefit of expert elicitation to develop a fuzzy-based FTA for leak assessment in subsea oil and gas production systems. The authors identified critical areas of leak initiation in subsea oil and gas production systems. These include pipelines, critical facilities, gas and oil wells, and third-party damage. The developed model identified the interactive structure among risk factors based on necessary measures analysis. The improved FTA have shown the different level of dependency modeling in complex engineering systems [14–18]. However, the instability in the failure causative factors and their non-linearity provide the basis for continued improvement for a robust failure model that can capture stochastic interactions.

6.2 The Complexity of Subsea Components Failure

The failure mechanism of subsea components exhibits complexity due to unstable dependence on the influencing factors. The failure could be instigated by corrosion, fatigue, structural damage, etc. [19–21]. These mechanisms created an interconnective structure that may be linear or non-linear. The structural configuration exhibits a different level of impact on the failure prediction precision in harsh offshore environments. The domino-based influence in the failure propagation further complicates the assessment process, especially in complex subsystems, e.g., BOP. The terrain-specific challenges of subsea operations have resulted in the premature failure of critical subsea facilities. In the study of [7, 20, 22, 23] further explored the complexity of the behavior of the subsea system under multiple failures. The authors applied the Important Sampling methodology to capture two competing failure modes due to corrosion. The small leak and burst failure functional parameters exhibit dependence and are randomly stochastic. The weighted importance sampling density for the completing failure modes was established to define the failure domains. Cai et al. [24] evaluated the performance of the subsea blowout preventer (BOP) to identify the common cause of failure using Markovian stochastic techniques. The authors split the BOP system into modules due to its complexity and evaluated the performance rating of the two-component subsea BOP design. BOP is a safety device used in offshore drilling operations to prevent uncontrolled releases of oil or gas from a well. It is typically located on the sea floor and is connected to the drilling rig by a series of pipes and hoses. The BOP is designed to seal the wellbore and prevent the uncontrolled flow of oil or gas in the event of a blowout or a sudden and uncontrollable release of fluids from a well. It consists of several components, including hydraulically operated rams that can close around the drill pipe and seal off the wellbore and a set of shear rams that can sever the drill pipe in case of an emergency. This considers the effect of stack configurations and mounting types of subsea control pods on the system performance.

The complex subsea BOP has two central systems: the subsea BOP stack and the control system. The formal is equipped with an LMRP connecter, two hydraulic connectors, and a wellhead connector. The functions of these elements are interconnected as well. The wellhead connector connects the BOP stack to the well, and the LMRP is connected by two annular preventers [11, 22]. The latter, which is the subsea BOP control system, consists of the surface and subsea components. The surface comprises the central control unit (CCU), which performs functional and pressure regulation capability. The elements of the CCU are redundantly configured based on triple modular redundancy configurations [25]. The different subsystems of the CCU interact with other systems based on their structural performance. The performance of these components influences the overall BOP reliability. The level of interaction or co-dependencies of the subsea BOP components will determine the reliability of the failure prediction in harsh offshore operations. It is important to note that the degree of stochastic interactions can be defined by the correlation parameters based on the subsystem's failure data. The following section explores recent studies to adequately capture the complexity and the non-linearity in the complex subsea system failure assessment.

6.3 Recent Advances in Subsea Components Failure Assessment

There have been advances in the design of subsea systems toward safety in harsh offshore environments for oil and gas operations. The recent design presents different complexity and safety concerns that are not comprehensively understood. Moreno-Trejo and Markeset [26] further suggested that the improved technological solutions may increase the uncertainty and unforeseen events due to limited testing and performance qualification. Several failure mechanisms in subsea systems; include mechanical failure, material defects, vortex-induced vibrations, corrosion, sand erosion, cavitation erosion, cracking, mechanical wear, fatigue, equipment plugging, fouling, flashing, deposition of wax, and droplet erosion. The prediction of failure of the advanced subsea system design under data uncertainty and operational complexity required a robust and integrated approach. Specialized subsea systems are employed at each subsea field development phase. There is a need to capture the key influencing factors in subsea design through a structural methodology, especially in harsh arctic deep and ultra-deep-water operations. There may be domain-specific constraints that need to be explored to improve the failure prediction strategy in critical subsea operations. The current study explored two critical subsea systems and their failure prediction mechanisms (subsea blowout preventer and subsea pipelines).

6.3.1 Subsea Blowout Preventer (SBOP)

The multi-structural features of the SBOP (can be find in Ref. [11]) and their performance dependency complicate the failure prediction and integrity management of the SBOP. To predict the system reliability under the structural dependency of the SBOP elements requires a proper understanding of the functional interactions among subsystems and their passivity. There has been an improved redundancy toward failure reduction in SBOP operations. The design SBOP test ram has the potency to reduce the likelihood of closure of the BOP. In the enhanced design, two sealing shear rams are appropriate due to drilling margin issues and loss of position risk for profound and ultra-deep water operations [27].

For instance, Picha et al. [28] studied the technological advancement and reliability concern for deep water subsea blowout preventers. The authors identify the complexity as a factor that limits reliable failure prediction of the BOP. Despite the application of the failure-based models, such as the Reliability Block Diagrams, FTA, and FMEA, to understand the failure propagation process, the control systems, and the choke and kill lines still experience a high degree of failure [26, 29, 30]. The need to understand the complexity of subsea components (e.g., a subsea blowout preventer (SBOP)) design and operations have informed the development of improved methods. Liu and Liu [31] studied the reliability of the BOP control system, considering the complex interactions among the elements during operations. The authors structurally capture the interactions through an adaptive probabilistic method, the Bayesian network (BN). The BN provides an updating capability to model scenarios with incomplete data to predict the failure profile of the complex SBOP systems.

Machado et al. [32] applied machine learning and artificial intelligence to explore the operational condition of the SBOP components toward condition-based maintenance. The generated historical data during the BOP operations were collected to develop a standard system behavior using a machine learning algorithm. The predicted deviations provide information that identifies the failure's root causes and retro-feed them into the learning mechanism for future failure trend prediction. Meng et al. [33] proposed an integrated system theoretic accident model and process (STAMP) and BN model for the safety assessment of SBOP. The STAMP explored the failure scenarios through a hierarchical structure to understand the propagation of the failure. The structure is mapped into the BN for probability prediction for the identified failure modes. The approach reliably evaluates the components' interactions and captures the structural modeling's complexity.

6.3.2   Subsea Pipelines

Subsea field development types are dependent on the geological characteristics of the reservoir. Understanding the reservoir features will inform the necessary subsea infrastructures and total oil and gas assets' value, including the expected year of production [26]. The oil and gas field features also affect the design features of the selected pipeline infrastructures. The criticality of the performance and integrity of the subsea assets toward safe operation in harsh environments needs to be explored. A robust risk-based failure assessment framework can predict the complexity of the subsea infrastructures and their performances.

Subsea pipelines serve several purposes in subsea field development and offshore operations. The subsea pipeline system can be a pipe-in-pipe, single-pipe pipeline, or bundle system. Most time, the subsea pipelines are called subsea flowlines, depending on their functionality. The reliability of these pipelines is crucial to sustainable offshore operations in harsh environments. Several studies that explore various risk factors for the reliability prediction of oil and gas pipelines are presented in the literature [12, 34–43].

6.3.2.1   Pipeline Failure Assessment Approaches

To further understand the pipeline failure prediction approaches [34], reviewed the various failure prediction models for offshore and onshore oil and gas systems. The author examined the quantitative and qualitative methods used to assess system failures and the associated risk. These methods include the FTA, ETA, fuzzy-logic-based failure assessment, FMEA, bow-tie graphical model, Bayesian network, index-based method, and matrix-based method.

Index-based risk management explored the semi-quantitative methods for oil and gas pipelines. The index risk assessment approach integrates the causative factors, inherent risk, and consequences with associated weight [40, 44–46]. The relative risk score is used as indices for the failure prediction of the subsea pipeline. For example, Khaleghi et al. [47] explore the failure of a gas pipeline using the relative risk scores through an aggregated risk index. The matrix-based risk assessment approach explores the probability and consequences of failure based on ranking criteria. Several statistical analyses and theoretical, quantitative models to estimate the risk of failures in various oil and gas pipelines based on the matrix-based configuration are presented in the referenced literature [34, 48–50]. Jamshidi et al. [50] proposed a new fuzzy inference system for pipeline risk assessment. The authors used a relative risk score approach as an intelligent tool for oil and gas pipeline risk prediction. Shahriar et al. [51] integrated fuzzy logic with bow-tie analysis. The authors proposed the fuzzy utility value based on triple-bottom-line sustainability criteria for natural gas pipeline failure risk assessment under uncertainty. Jianxing et al. [52] explore the application of the weakest t-norms-based fuzzy fault tree analysis for submarine pipeline failure risk prediction. The authors evaluate the reliability of the weakest t-norms comparatively approach in the reliable prediction of the probability of failure based on expert knowledge. Further applications of the fuzzy-based have been explored by various researchers for pipeline failure risk predictions [12, 17, 53, 54].

6.3.2.2   Dynamic and Data-Driven Approaches for Subsea Components Failure Assessment

The complex interactions among failure-influencing parameters required a dynamic approach. Several studies have proposed dynamic risk assessment methods for subsea components failure risk prediction. Kabir et al. [55] present a fuzzy Bayesian belief network for the safety assessment of oil and gas pipelines. The model explicitly captures the dependencies and updates the probability of the failure of causal factors. Shan et al. [56] explored the integration of the fuzzy-Bayesian Bow-tie approach for failure risk prediction of a natural gas pipeline. The approach captures the uncertainty associated with the failure influencing factors to identify their critical impact on the failure risk. Yuanjiang et al. [9] present a failure risk assessment model based on a dynamic Bayesian network. The authors diagnostically explored the dynamic fatigue failure probability of the subsea wellheads over time to identify the most inferential risk factors.

Further analysis that presents a robust integrated model for the subsea component has been demonstrated in the work of Singh et al. [57]. The authors proposed a novel approach that integrates the application of the α-cut interval-based fuzzy fault tree with a Bayesian network for submarine pipeline leak failure analysis. The approach employed the weakest t-norm and α cut-based similarity aggregation tool to predict the initial probabilities of the primary events. These probabilities are then updated using a fuzzy Bayesian network. Further applications of dynamic Bayesian networks for subsea components failure risk prediction have been demonstrated in the referenced studies [21, 26, 58, 59].

The increase in data generation and functional complexity of subsea systems have informed the need to explore a data-driven model for failure risk prediction. This also supports the need to digitize the subsea process in the industry 4.0 environment. Heidary et al. [60] explored various data-driven approaches for the failure prediction of oil and gas pipelines. The study examined the potency of the probabilistic data-driven failure models based on their applicability. Zakikhani et al. [61] developed three objective data-driven models based on artificial neural network (ANN) and multinominal logit (MNL) regression for failure assessment of oil pipelines. The approach explores the various failure mechanism of oil and gas pipelines, such as mechanical failure, corrosion-induced failure, and third-party failure. The model reliably predicts the failure index with an average validity percentage (AVP) of 72.8%. Li et al. [62] explored the potency of a data-driven machine-learning model for subsea system failure prediction in an industry 4.0 environment. The authors hybridized three data-driven methods (principal component analysis, artificial bee colony, and support vector regression) for corrosion-induced failure prediction of subsea pipelines. The approach shows feasibility and effectiveness with high accuracy and robustness.

Eastvedt et al. [63] proposed the application of a regression-supervised machine learning algorithm for subsea pipeline fault-based failure analysis. The authors used computational fluid dynamics (CFD) to train the algorithm to generate a dataset based on the flow parameters. The model offers a proactive monitoring capacity for subsea pipeline networks against failure in the harsh ocean environment. Eastvedt et al. [63] explored the integration of a radial basis function, multilayer perception neural network, and multinomial logistic regression for failure analysis of oil and gas pipelines. The authors used historical data to develop a tool to detect pipeline failure sources. To further explore the data complexity modeling for subsea operations, Elshaboury et al. [64] applied a radial basis function (RBF) neural network, multilayer perceptron (MLP) neural network, and a multinomial logistic (MNL) regression on subsea pipeline operational data. The modeling seeks to validate the performance of the selected algorithms on the data to learn the failure pattern of the subsea system given complex interdependencies. The model captures the triggering events that influence the failure mechanism of subsea pipelines and identifies the average validity of 84%, 85%, and 81% for the MLP, RBF, and MNL, respectively. Figure 6.1 shows the trend of data-driven machine learning (DDML) models that are applied in the subsea system operational data analysis for failure pattern prediction and condition monitoring. These model frameworks could be mapped as an integral part of data twin structure and digitalization processes in the industry 4.0 environment for robust and reliable failure pattern prediction in subsea system operations.

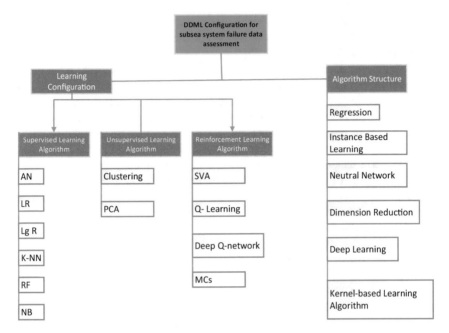

Fig. 6.1 A classification of DDML for subsea system failure analysis. where AN—Artificial neural network, LR—Linear Regression, LgR—Logistic Regression, K-N—K Nearest Neighbour, SVM—Support Vector Machine, RF—Random Forest, NR—Naïve Bayes, PCA—Principal Component Analysis, SVA—Singular Value Analysis, MCs—Monte Carlo simulation

6.4   Further Considerations

The subsea systems and their operating environment pose complexity in the modeling characterization, especially in a harsh and unstable environment. The failure mechanisms of various subsea components are influenced by multiple environmental and operational factors that exhibit complex interdependencies. Such complexity needs to be captured in a reliable failure risk prediction. The data-driven machine learning model has shown the capacity for subsea components' short- and long-term-failure prediction. Nevertheless, limited consideration has been given to exploring the concept of multi-hazard events in the modeling framework for failure risk indicators. There is a need to develop a trustworthiness criterion for a multi-hazard risk aggregation structure for subsea system operations in harsh environments. The study identified areas for further research in subsea system failure prediction. These include.

i.   The development of risk-based integrated models capturing data interdependencies and risk trustworthiness
ii. The development of data-driven advanced theories and frameworks via data digitalization and IoT for subsea system management in harsh environments.

iii. Advanced resilient design for structural sustainability of subsea components in harsh ocean environments.

iv. The development of an integrated DDML with multi-hazard risk aggregation for failure monitoring of critical subsea components in harsh offshore operations.

6.5 Conclusions

This chapter presents the state of knowledge on failure assessment approaches for subsea components in harsh offshore operations. The study takes into account the fundamental theory of failure and risk assessments and their progressive improvement for better prediction. For example, various qualitative approaches for failure risk prediction, such as FMEA, and risk matrix for subsea system applications, were reviewed. The merits of quantitative modeling, representing causal interconnected among the subsea subsystems' functionality, were reviewed. The study further explained the potency of the data-driven machine learning algorithms in failure risk prediction of subsea components in harsh offshore environments. Knowledge gaps for further research were identified. There is a need for continued research to develop integrated data-driven machine learning multi-hazard risk aggregation tools to capture the complex interdependencies among system functionality in subsea field operations. The integrated algorithms will be able to capture the instability in the escalating failure risks factors, such as human errors, natural disasters, environmental factors, operational parameters, and met ocean parameters. This can be mapped with safety criteria to explore the varying degrees of uncertainty in system performance for a well-informed failure management strategy of subsea components in harsh offshore operations.

References

1. Adumene, S., Khan, F., Adedigba, S., Mamudu, A., Rosli, M.I.: Offshore oil and gas development in remote harsh environments: engineering challenges and research opportunities. Saf. Extrem. Environ. 0123456789 (2022. https://doi.org/10.1007/s42797-022-00057-1.
2. Necci, A., Tarantola, S., Vamanu, B., Krausmann, E., Ponte, L.: Lessons learned from offshore oil and gas incidents in the Arctic and other ice-prone seas. Ocean Eng. **185**(May), 12–26 (2019). https://doi.org/10.1016/j.oceaneng.2019.05.021
3. Yazdi, M., Adesina, K.A., Korhan, O., Nikfar, F.: Learning from fire accident at Bouali Sina petrochemical complex plant. J. Fail. Anal. Prev. (2019). https://doi.org/10.1007/s11668-019-00769-w
4. N.G. and B. Brazilian National Agency of Petroleum: Investigation Report of the 11/02/2015 in the Fpso Cidade De São Mateus Superintendence of Operational Safety and the Environment (SSM) (2015)
5. Kvitrud, A., Løland, A.H.: Observed wave actions on Norwegian semi-submersible and TLP decks (2018)

6. Adumene, S., Okwu, M., Yazdi, M., Afenyo, M., Islam, R., Orji, C.U., Obeng, F., Goerlandt, F.: Dynamic logistics disruption risk model for offshore supply vessel operations in Arctic waters. Marit. Transp. Res. **2**, 100039 (2021). https://doi.org/10.1016/j.martra.2021.100039
7. Gupta, J., Talukdar, M.K., Velusshami, S.K., Sharma, A., Makkar, S.: Premature failure of submarine well fluid lines: a case study. J. Fail. Anal. Prev. **21**(2), 363–369 (2021). https://doi.org/10.1007/s11668-020-01051-0
8. Chen, Y., et al.: Collapse failure and capacity of subsea pipelines with complex corrosion defects. Eng. Fail. Anal. **123**(March 2020), 105266 (2021). https://doi.org/10.1016/j.engfailanal.2021.105266
9. Yuanjiang, C., et al.: Dynamic Bayesian networks based approach for risk analysis of subsea wellhead fatigue failure during service life. Reliab. Eng. Syst. Saf. **188**(March), 454–462 (2019). https://doi.org/10.1016/j.ress.2019.03.040
10. Drumond, G.P., Pasqualino, I.P., Pinheiro, B.C., Estefen, S.F.: Pipelines, risers and umbilicals failures: A literature review. Ocean Eng. **148**(November 2017), 412–425 (2018). https://doi.org/10.1016/j.oceaneng.2017.11.035
11. Shafiee, M., Enjema, E., Kolios, A.: An integrated FTA-FMEA model for risk analysis of engineering systems: a case study of subsea blowout preventers. Appl. Sci. **9**(6), (2019). https://doi.org/10.3390/app9061192
12. Yuhua, D., Datao, Y.: Estimation of failure probability of oil and gas transmission pipelines by fuzzy fault tree analysis. J. Loss Prev. Process Ind. **18**(2), 83–88 (2005). https://doi.org/10.1016/j.jlp.2004.12.003
13. Cheliyan, A.S., Bhattacharyya, S.K.: Fuzzy fault tree analysis of oil and gas leakage in subsea production systems. J. Ocean Eng. Sci. **3**(1), 38–48 (2017). https://doi.org/10.1016/j.joes.2017.11.005
14. Kabir, S.: An overview of fault tree analysis and its application in model based dependability analysis. Expert Syst. Appl. **77**, 114–135 (2017). https://doi.org/10.1016/j.eswa.2017.01.058
15. Gholamizadeh, K., Zarei, E., Omidvar, M., Yazdi, M.: Fuzzy sets theory and human reliability: review, applications, and contributions BT—Linguistic methods under fuzzy information in system safety and reliability analysis, pp. 91–137. In: Yazdi, M. (ed.) Springer, Cham (2022)
16. Kabir, S., Geok, T.A.N.K.I.M., Kumar, M., Yazdi, M., Hossain, F.: A method for temporal fault tree analysis using intuitionistic fuzzy set and expert elicitation. IEEE Access **8**, 980–996 (2020)
17. Kabir, S., Yazdi, M., Aizpurua, J.I., Papadopoulos, Y.: Uncertainty-aware dynamic reliability analysis framework for complex systems. IEEE Access 6 (2018)
18. Mohammadfam, I., Zarei, E., Yazdi, M., Gholamizadeh, K.: Quantitative risk analysis on rail transportation of hazardous materials. Math. Probl. Eng. **2022**, 6162829 (2022)
19. Fu, G., Yang, W., Li, C., Shi, W.: Reliability analysis of corrosion affected underground steel pipes considering multiple failure modes and their stochastic correlations. Tunn. Undergr. Sp. Technol. **87**(February), 56–63 (2019). https://doi.org/10.1016/j.tust.2019.02.005
20. Gong, C., Zhou, W.: Importance sampling-based system reliability analysis of corroding pipelines considering multiple failure modes. Reliab. Eng. Syst. Saf. **169**(August 2017), 199–208 (2018). https://doi.org/10.1016/j.ress.2017.08.023
21. Adumene, V., Khan, F., Adedigba, S., Zendehboudi, S.: Offshore system safety and reliability considering microbial influenced multiple failure modes and their interdependencies. Reliab. Eng. Syst. Saf. 215(November), 107862 (2021). https://doi.org/10.1016/j.ress.2021.107862
22. Yazdi, M., Adumene, S., Zarei, E.: Introducing a probabilistic-based hybrid model (fuzzy-BWM-Bayesian network) to assess the quality index of a medical service BT—Linguistic methods under fuzzy information in system safety and reliability analysis. In: Yazdi, M. (ed.) Springer, Cham, pp. 171–183 (2022).
23. Adumene, S., Adedigba, S., Khan, F., Zendehboudi, S.: An integrated dynamic failure assessment model for offshore components under microbiologically influenced corrosion. Ocean Eng. **218** (2020). https://doi.org/10.1016/j.oceaneng.2020.108082
24. Cai, B., Liu, Y., Liu, Z., Tian, X., Zhang, Y., Liu, J.: Performance evaluation of subsea blowout preventer systems with common-cause failures. J. Pet. Sci. Eng. **90–91**, 18–25 (2012). https://doi.org/10.1016/j.petrol.2012.04.007

25. Shafiee, M., Elusakin, T., Enjema, E.: Subsea blowout preventer (BOP): design, reliability, testing, deployment, and operation and maintenance challenges. J. Loss Prev. Process Ind. 66(May) (2020). https://doi.org/10.1016/j.jlp.2020.104170.
26. Moreno-Trejo, J., Markeset, T.: Identifying challenges in the development of subsea petroleum production systems. IFIP Adv. Inf. Commun. Technol. November, 287–295 (2012). https://doi.org/10.1007/978-3-642-33980-6
27. Holand, P., Awan, H.: Reliability of deepwater subsea BOP systems and well kicks (2012)
28. Picha, M.S., Abdullah, T.M., Rai, A., Sinha, S., Patil, P.A.: Deepwater subsea BOP technological and reliability advancement. In: International Petroleum Technology Conference- IPTC 21430-mS, no. April, pp. 1–16 (2021). https://doi.org/10.2523/iptc-21430-ms
29. Schneider, K.D.: Risk and reliability analysis of a subsea system for oil production December 2018 (2018)
30. dos Reis Costeira, M.J.: Reliability modelling of subsea production equipment (2011)
31. Z. Liu and Y. Liu, "A Bayesian network based method for reliability analysis of subsea blowout preventer control system," J. Loss Prev. Process Ind., vol. 59, no. December 2018, pp. 44–53, 2019, doi: https://doi.org/10.1016/j.jlp.2019.03.004.
32. Machado, R.C., Leite, F., Xavier, C., Albuquerque, A., Lima, S., Carvalho, .: Development of failure prediction models for subsea blowout preventers using data analytics and AI (2021). Available https://doi.org/10.4043/31027-MS
33. Meng, X., Chen, G., Zhu, J., Li, T.: Application of integrated STAMP-BN in safety analysis of subsea blowout preventer. Ocean Eng. 258(September 2020), 111740 (2022). https://doi.org/10.1016/j.oceaneng.2022.111740
34. Zakikhani, K., Nasiri, F., Zayed, T.: A review of failure prediction models for oil and gas pipelines. J. Pipeline Syst. Eng. Pract. 11(1) (2020). https://doi.org/10.1061/(asce)ps.1949-1204.0000407
35. Li, X., Han, Z., Yazdi, M., Chen, G.: A CRITIC-VIKOR based robust approach to support risk management of subsea pipelines. Appl. Ocean Res. 124, 103187 (2022). https://doi.org/10.1016/j.apor.2022.103187
36. Yazdi, M., Khan, F., Abbassi, R., Quddus, N., Castaneda-Lopez, H.: A review of risk-based decision-making models for microbiologically influenced corrosion (MIC) in offshore pipelines. Reliab. Eng. Syst. Saf. 223(February), 108474 (2022). https://doi.org/10.1016/j.ress.2022.108474
37. Liu, A., Chen, K., Huang, X., Chen, J., Zhou, J., Xu, W.: Corrosion failure probability analysis of buried gas pipelines based on subset simulation. J. Loss Prev. Process Ind. 57(August 2018), 25–33 (2019). https://doi.org/10.1016/j.jlp.2018.11.008
38. Chandrasekaran, S.: Offshore structural engineering: reliability and risk assessment. CRC Press Taylor & Francis Group (2016)
39. Gomes, W.J.S., Beck, A.T.: Optimal inspection and design of onshore pipelines under external corrosion process. Struct. Saf. 47, 48–58 (2014). https://doi.org/10.1016/j.strusafe.2013.11.001
40. Li, X., Chen, G., Zhu, H.: Quantitative risk analysis on leakage failure of submarine oil and gas pipelines using Bayesian network. Process Saf. Environ. Prot. 103, 163–173 (2016). https://doi.org/10.1016/j.psep.2016.06.006
41. Meng, X., Chen, G., Zhu, G., Zhu, Y.: Dynamic quantitative risk assessment of accidents induced by leakage on offshore platforms using DEMATEL-BN. Int. J. Nav. Archit. Ocean Eng. (2018). https://doi.org/10.1016/j.ijnaoe.2017.12.001
42. Cowin, T.G., Lanan, G.A., Paulin, M., DeGeer, D.: Integrity monitoring of offshore arctic pipelines. In: Proceedings of International Conference Offshore Mechanics and Arctic Engineering—OMAE, vol. 4, pp. 1–11 (2021). https://doi.org/10.1115/OMAE2021-64174
43. Li, X., Chen, G., Chang, Y., Xu, C.: Risk-based operation safety analysis during maintenance activities of subsea pipelines. Process Saf. Environ. Prot. 122, 247–262 (2019). https://doi.org/10.1016/j.psep.2018.12.006
44. Li, H., Yazdi, M., Huang, H.-Z., Huang, C.-G., Peng, W., Nedjati, A., Adesina, K.A.: A fuzzy rough copula Bayesian network model for solving complex hospital service quality assessment. Complex Intell. Syst. (2023). https://doi.org/10.1007/s40747-023-01002-w.

45. Yazdi, M., Khan, F., Abbassi, R., Rusli, R.: Improved DEMATEL methodology for effective safety management decision-making. Saf. Sci. **127**, 104705 (2020). https://doi.org/10.1016/j.ssci.2020.104705
46. Xie, M., Tian, Z.: A review on pipeline integrity management utilizing in-line inspection data. Eng. Fail. Anal. **92**(May), 222–239 (2018). https://doi.org/10.1016/j.engfailanal.2018.05.010
47. Yazdi, M., Mohammadpour, J., Li, H., Huang, H.-Z., Zarei, E., Pirbalouti, R.G., Adumene, S.: Fault tree analysis improvements: a bibliometric analysis and literature review. Qual. Reliab. Eng. Int. n/a. (2023). https://doi.org/10.1002/qre.3271
48. Alxxxjaroudi, S., Ulxxxhamid, A., Alxxxgahtani, M.M.: Failure of crude oil pipeline due to microbiologically induced corrosion. Corros. Eng. Sci. Technol. **46**(4), 568–579 (2011). https://doi.org/10.1179/147842210X12695149033819
49. Aljaroudi, A., Khan, F., Akinturk, A., Haddara, M.: Risk assessment of offshore crude oil pipeline failure. J. Loss Prev. Process Ind. **37**, 101–109 (2015). https://doi.org/10.1016/j.jlp.2015.07.004
50. Jamshidi, A., Yazdani-Chamzini, A., Yakhchali, S.H., Khaleghi, S.: Developing a new fuzzy inference system for pipeline risk assessment. J. Loss Prev. Process Ind. **26**(1), 197–208 (2013). https://doi.org/10.1016/j.jlp.2012.10.010
51. Shahriar, A., Sadiq, R., Tesfamariam, S.: Risk analysis for oil & gas pipelines : a sustainability assessment approach using fuzzy based bow-tie analysis (October 2017) (2012). https://doi.org/10.1016/j.jlp.2011.12.007
52. Jianxing, Y., Haicheng, C., Yang, Y., Zhenglong, Y.: A weakest t-norm based fuzzy fault tree approach for leakage risk assessment of submarine pipeline. J. Loss Prev. Process Ind. **62**(135), 103968 (2019). https://doi.org/10.1016/j.jlp.2019.103968
53. Yang, M., Khan, F.I., Sadiq, R.: Prioritization of environmental issues in offshore oil and gas operations: a hybrid approach using fuzzy inference system and fuzzy analytic hierarchy process. Process Saf. Environ. Prot. **89**(1), 22–34 (2011). https://doi.org/10.1016/j.psep.2010.08.006
54. Badida, P., Balasubramaniam, Y., Jayaprakash, J.: Risk evaluation of oil and natural gas pipelines due to natural hazards using fuzzy fault tree analysis. J. Nat. Gas Sci. Eng. **66**(January), 284–292 (2019). https://doi.org/10.1016/j.jngse.2019.04.010
55. Kabir, G., Sadiq, R., Tesfamariam, S.: A fuzzy Bayesian belief network for safety assessment of oil and gas pipelines. Struct. Infrastruct. Eng. **12**(8), 874–889 (2016). https://doi.org/10.1080/15732479.2015.1053093
56. Shan, X., Liu, K., Sun, P.: Risk analysis on leakage failure of natural gas pipelines by fuzzy bayesian network with a Bow-Tie model. **2017**(2) (2017)
57. Singh, K., Kaushik, M., Kumar, M.: Integrating α-cut interval based fuzzy fault tree analysis with Bayesian network for criticality analysis of submarine pipeline leakage: a novel approach. Process Saf. Environ. Prot. **166**(August), 189–201 (2022). https://doi.org/10.1016/j.psep.2022.07.058
58. Adumene, S., Khan, F., Adedigba, S., Zendehboudi, S., Shiri, H.: Dynamic risk analysis of marine and offshore systems suffering microbial induced stochastic degradation. Reliab. Eng. Syst. Saf. **207**(March 2021), 107388 (2021). https://doi.org/10.1016/j.ress.2020.107388
59. Yazdi, M., Khan, F., Abbassi, R., Quddus, N.: Resilience assessment of a subsea pipeline using dynamic Bayesian network. J. Pipeline Sci. Eng. **2**(2), 100053 (2022). https://doi.org/10.1016/j.jpse.2022.100053
60. Heidary, R., Gabriel, S.A., Modarres, M., Groth, K.M., Vahdati, N.: A review of data-driven oil and gas pipeline pitting corrosion growth models applicable for prognostic and health management. Int. J. Progn. Heal. Manage. 1–13 (2018)
61. Zakikhani, K., Zayed, T., Abdrabou, B., Senouci, A.: Modeling failure of oil pipelines. J. Perform. Constr. Facil. **34**(1), 1–10 (2020). https://doi.org/10.1061/(asce)cf.1943-5509.0001368
62. Li, X., Zhang, L., Khan, F., Han, Z.: A data-driven corrosion prediction model to support digitization of subsea operations. Process Saf. Environ. Prot. **153**, 413–421 (2021). https://doi.org/10.1016/j.psep.2021.07.031

63. Eastvedt, D., Naterer, G., Duan, X.: Detection of faults in subsea pipelines by flow monitoring with regression supervised machine learning. Process Saf. Environ. Prot. **161**, 409–420 (2022). https://doi.org/10.1016/j.psep.2022.03.049

64. Elshaboury, N., Alxxxsakkaf, A., Alfalah, G., Abdelkader, E.M.: Data-driven models for forecasting failure modes in oil and gas pipes. Processes **10**(2), 1–17 (2022). https://doi.org/10.3390/pr10020400

Chapter 7
An Intelligent Cost-Based Consequence Model for Offshore Systems in Harsh Environments

Abstract The present study proposed the application of an integrated probabilistic model for the failure consequence assessment of oil and gas pipelines suffering under-deposit corrosion (UDC). The physics of UDC potential based on the degradation mechanism is explored and built into a network to predict the corrosion rate. The predicted corrosion rate under the prevailing corrosion mechanism is mapped into a probabilistic structure to capture their interactions on the failure state of the asset. In this study, two-level of consequences are explored for the under-deposit corrosion mechanism failure-induced prediction. The proposed approach integrates the Bayesian network with the expected utility decision theory for the cost-based consequence analysis. The dynamic and updating capacity of the integrated approach provides an intelligent monitoring advantage for asset cost-based integrity management. The model was tested on oil transmission pipelines to determine the likely consequences of failure in terms of financial losses. The result shows for the worst-case scenario, the cost values of the consequence of failure give 1.824×10^7 USD, 1.245×10^5 USD, and 1.082×10^8 USD for the loss of production, compensation, and environmental impact, respectively. The present approach offers a cost-based assessment tool that could guide a well-informed integrity management strategy for oil and gas operations in harsh environments.

Keywords Failure probability · Under-deposit corrosion · Expected utility decision theory · UDC rate · Bayesian network · Consequence

7.1 Introduction

Offshore pipelines are critical infrastructures that support oil and gas operations and transportation. They are exposed to harsh environmental conditions that could facilitate their susceptibility to corrosion and failure over time [1]. The nature of the environmental and operational factors could further enhance the high rate of degradation of the steel pipelines. Depending on their potential impact and mode, these degradation-instigating factors can be classified into organic and inorganic. The associated cost of the failure of oil and gas pipelines due to corrosion is still a concern across many industries. It has been reported that corrosion failure accounted for about 20% of the major pipeline accidents in the United States [2–4]. This failure resulted in an estimated loss of about \$203 million [5]. Corrosion prevention and management performance are a function of properly understanding the influential factors' characteristics and the corrosion mechanism. Thus, taking into consideration the complex nature of corrosion, especially under-deposit corrosion, is key to ensuring the safety of pipeline operations.

The development process of under-deposit corrosion (UDC) in oil and gas pipeline are influenced by these organic and inorganic factors. Most commonly, the transporting fluid through the pipeline carries particles. Under a conducive environment, these particles gather and form deposits of different porosity at the six o'clock position in the pipeline. For gas pipelines, the water droplet or condensation provides a supportive environment for the deposit. As the environment becomes supportive, the deposit creates an isolated localized microenvironment with characteristics vastly different from the transporting fluid. Also, corrosion products, such as metallic products, are collected and could increase the rate of UDC. The deposits are further classified based on their sources, porosity, and fluid type. For instance, Inert deposit describes deposits such as clay, silica, and sand carried from the reservoir during the drilling process. The Active deposits are commonly described as metallic corrosion products from the steel pipe surface and could be transported with the media until they are deposited. Their characteristic impact increases the corrosion rate under UDC. Biofilm deposits, wax, and asphaltenes are organic deposits that could support corrosion and threaten the operational sustainability of the pipeline. It became complicated in a scenario where there are multispecies bacteria formation and deposits. The deposit may provide a sustainable environment for the biofilm degradation potential over long-term exposure.

Furthermore, the non-conducive deposit, such as silt, sand, and certain scales on the pipeline, can act as a diffusion barrier to limit the effectiveness of the applied biocides from the bulk solution to the metal surface. The interactions among other influential factors, such as CO_2 and H_2S, can further increase the UDC rate [6]. Huang et al. [7] examined the impact of inert solid deposits with different porosity on the CO_2 corrosion mechanism. The authors developed a linear relationship between the deposit porosity and the average CO_2 corrosion rate of the API X65 pipeline. The results indicate an increase in the corrosion rate as the deposit porosity increases. In the recent work of [8], the authors explored the impact of Iron Sulfide deposits on an X65 pipeline based on different particle sizes and crystalline structures. In 100-h autoclave corrosion testing, the mass loss specimens were fully covered, indicating that the deposit facilitated the corrosion mechanism.

Peacock and Grauman [9] further explored the steel material response to under-deposit corrosion in harsh environments. The authors studied the enhancing effect of resistance alloy on steel performance in harsh under-deposit corrosion-infested environments. The steel pipelines show different corrosion mechanisms and rates at different pH values. It was observed that the failure-induced characteristic of the under-deposit corrosion presents complexity, making it different to monitor [10]. The prevailing unstable environment is influenced by many variables that exhibit different levels of dependencies. This tendency is propagated to the failure mechanism of the asset. Gupta et al. [11] examined the failure of two well-fluid lines considering under-deposit internal corrosion. The authors identified that the fluid velocity, which was significantly low, enhanced the deposition settling at the 6 o'clock position in the pipeline. The study revealed that the likely interaction among the under-deposit, biofilm, and failed pigging process contributed to the premature failure of the well-fluid lines. Considering the significant impact of multivariate interactions on the under-deposit-induced failure of oil and gas pipelines is essential. The corrosion response variables are randomly characterized and need to be captured for reliable failure mechanism prediction for gas pipelines suffering UDC.

This chapter explores the application of a probabilistic model for failure consequence prediction of oil and gas pipelines suffering UDC. The mechanistic structure describes the physics of UDC formation and their influencing factors to predict the UDC rate. The predicted characteristics of the corrosion mechanism serve as input for the leak failure probability prediction under different risk factors. A consequence model is built based on the expected utility decision theory for the impact prediction of system failure in terms of financial losses. The study provides an integrated and dynamic approach to the cost-based assessment of UDC-induced failure consequences. The approach was tested on an under-deposit-induced corroding pipeline to develop the failure and consequence profiles. The model offers a predictive tool for cost-based consequence assessment to aid the integrity management of critical assets in oil and gas operations.

7.2 Dynamic Failure Assessment of Offshore System

Multiple influential factors, such as microbial species, deposits, fluid characteristics, operational parameters, and environmental parameters, instigate critical offshore infrastructure failure. The complexity of the failure mechanism, especially in the harsh offshore environment, depends on the degree of interdependencies of these risk factors. For instance, the under-deposit corrosion mechanism is influenced or supported by microbes, such as sulfur-reducing bacteria. The deterioration process can worsen, resulting in premature failure of the critical assets. Assessing the failure-instigating potential of this corrosion mechanism requires a robust dynamic method. A method that can capture the complex interaction and integration of the various risk factors to predict the propagation of the failure consequences in long-term operations.

7.2.1 Structural Learning Using Bayesian Network

The complex interaction among failure-instigating factors in offshore operations can be captured using dynamic tools. The essence is to learn from the data set to understand the system's behavior and define the pattern of failure for a given phenomenon.

The Bayesian network demonstrates the capacity to capture the multidimensionality in the data sets and their interactions for failure propagation prediction.

Bayesian networks: The Bayesian network (BN) is a technique that models the random variables under uncertainty using a directed acyclic graph structure [12]. BN modeling can identify both the qualitative and quantitative topologies, which in most cases are based on the d-separation notion and direct dependence among random variables. Its advantages range from its graphical decoding of conditional independence to joint probability representation among random variables [1].

For a given set of basic random variables $(Y_1, Y_2, \ldots, Y_n)$, Eq. (7.1) describes the joint conditional probability distribution, $P(U)$.

$$P(U) = P[Y_1, Y_2, Y_3, \ldots, Y_n] = \prod_{i=1}^{n} P[(Y_i | Parent(Y_i))] \qquad (7.1)$$

where $P(U)$ is the joint probability distribution, and $Parent(Y_i)$ is the parent of the set of random variables, Y_i.

The BN consists of various algorithms for inference and computing of the posterior probability distribution on a set of query variables designated as Q, for a given evidence called E (i.e., $P(Q|E)$). Figure 7.1 shows the BN structure for random variables X_1, X_2, X_3, X_4. Also, the BN updates the probabilities when new information about the variables is available. For example, given an observed variable X_3 to be in state e, the joint probability distribution can be updated based on Bayes' theorem, as shown by Eq. (7.2).

$$P(X_1, X_2, X_4 | e) = \frac{P(X_1, X_2, X_4 | e)}{\sum_{X_1, X_2, X_4} P(X_1, X_2, X_4, e)} \qquad (7.2)$$

Different structural learning algorithms define the interactions of key influencing variables for a BN structure. The score-based and constraints-based structural learning approaches are commonly used in building BN structures. These approaches can learn the dataset based on their conditional independence based on data points and maximize the likelihood of the model [13–16]. Effective learning algorithms, such as the PC and the Incremental Association Markov Blanket (IAMB), have

Fig. 7.1 BN learning structure

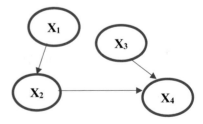

proven useful in the constraint-based learning of the structure. The maximum Likelihood Estimation (MLE) learns the structure based on a likelihood maximizer and is described by Eq. (7.3) [14].

$$\mathcal{LL}(\mathbb{G}; D) = \prod_{d \in D} P(d|\mathbb{G}) \qquad (7.3)$$

where $\mathcal{L}$ is the likelihood maximizer, D is a set of failure data, and $\mathbb{G}$ is a given model?

Furthermore, the Bayesian Information Criteria (BIC) adopts the log-likelihood and regularization terms to characterize the observed dataset for failure propagation. This concept is described by Eq. (7.4).

$$BIC(\mathbb{G}; D) = \mathcal{LL}(\mathbb{G}; D) - \frac{\log m}{2} dim(\mathbb{G}) \qquad (7.4)$$

where D denotes the dataset, m indicates the number of samples, and $dim(\mathbb{G})$ is the number of parameters in the BN model.

The relational framework is constructed considering the root nodes' likelihood of occurrence and the conditional probability, which describes the complex interactions. The characterized interactions tables are formulated from historical data, expert opinions, and literature.

7.2.2 Failure Consequence Analysis

The failure of oil and gas infrastructures present different level of consequences. The likely consequences impact humans, the environment, and systems. To quantify failure consequences, recent studies have proposed different methods [17–21]. To capture the multi-failure modes and failure impact classification, Adumene et al. [21] developed a two-step dynamic consequences assessment model for economic loss prediction under stochastic degradations. This approach is adopted and integrated with the UDC prediction mechanism developed in Sect. 2.1. Equation 7.5 is used to model the economic losses based on the two-step consequences and the expected utility decision theory.

Let the failure event be called B; the expected utility of the failure event, B, is defined as

$$EU(B) = \sum_{o \in O} P_B(o)U(o) \qquad (7.5)$$

where O is the set of outcomes, $P_B(o)$ is the probability of outcome o conditional on B and $U(o)$ is the utility of o. According to [22], $P_B(o)$ is the summation of the probabilities of the states that when combined with act B, lead to the outcome, o.

The overall analysis provides an integrated and proactive framework for consequences assessments considering the financial losses on investment and the environment.

7.3 Application and Results Analysis

The developed dynamic consequences assessment model is demonstrated on the subsea pipeline under complex degradation mechanisms. An API X60 transmission pipeline operating in a harsh environment with severe localized UDC is used [23–25]. The pipeline management strategy includes a 4.70 ppm dose of corrosion inhibitor. The investigation revealed severe localized defects at the 6 o'clock position of the pipeline. Fluid characteristics and microbe supported the complex failure mechanism of the pipeline. The cost parameters are selected after the works of [24–31]. The following subsections show the result analysis of the model application.

7.3.1 UDC Propagation Prediction Under Unstable Influencing Factors

The UDC presents an unstable propagation mechanism that requires a dynamic tool to adequately predict its degradation impact. The Bayesian network provides the capability to capture these influential factors for a holistic UDC prediction under uncertainty. Figure 7.2 shows the parametric learning of the BN structure for UDC rate prediction.

The result reveals that there is a likelihood of a 58% high rate of UDC for a 55% abundance of deposition parameters, 56% abundance of operating parameters, 60% abundance of fluid supporting characteristics, and 63% abundance of material parameters, respectively. The result further indicates the role of the identified intermediate influential parameters. Organic and inorganic parameters, including biofilm characteristics, influence the deposition parameters. The biofilm structure presents a more complex mechanism that is not explored in the current work. Further analysis is done by placing evidence on the Deposition parameters node to predict the degree of impact on the UDC rate. As shown, the UDC rate increased by 12.1%. Similarly, the UDC rate increases by 15.5, 22.4, and 6.9% for the operating parameters, material parameters, and fluid chemistry, respectively. For a 100% abundance of the UDC supporting parameters, as shown in Fig. 7.3, the high UDC rate increases the likelihood of 0.946.

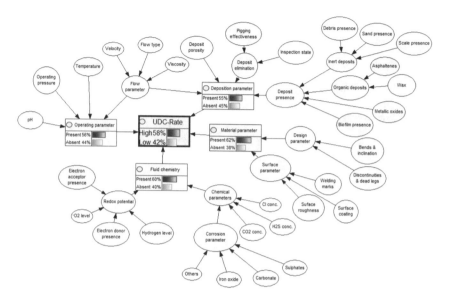

Fig. 7.2 Parametric learning of the BN structure for UDC rate prediction

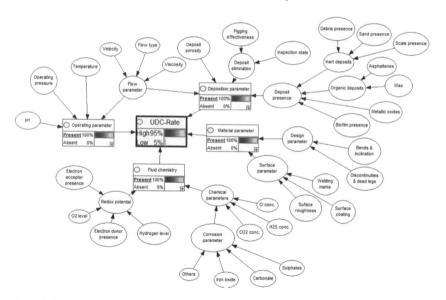

Fig. 7.3 BN structure learning for UDC rate prediction considering the evidence

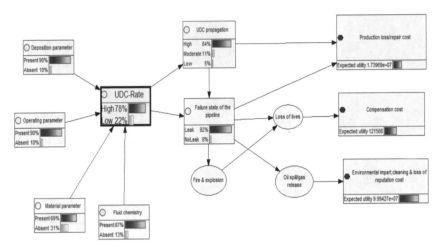

Fig. 7.4 BN structure learning for failure consequences assessment

7.3.2   An Integrated Cost-Based Consequences Assessment

The consequence assessment is quantified in financial terms based on the expected utility decision theory for different failure scenarios induced by the UDC. Figure 7.4 shows the learning of the developed dynamic structure for the cost-based consequence analysis.

The assessment is based on a two-steps consequences analysis where the oil spill is due to the UDC and is grouped into "Catastrophic oil spill (> 200,000 barrels)", "Major oil spill (10,000–200,000 barrels)", "Minor oil spill (< 10,000 barrels)", "No oil spill'. The result of the parametric learning of the intelligent model structure shows that at a UDC rate of 0.775, the likelihood of leak failure is 0.923. This indicates a high probability of failure at a severe UDC rate. The economic losses based on the expected utility decision theory at the leak failure probability are 1.73969×10^7 USD and 9.99427×10^7 USD for the loss of production/repair cost, environmental impact, clean up, and loss of reputation, respectively. The impact of the leak failure on the environment is quantified based on the cost of natural resources and their restoration from a moderate spill, as recommended [29].

To further assess the impact of the complex interaction of the key influential factors on the UDC rate and failure consequences, evidence is placed on the deposition parameter node, operating parameters node, material parameter node, and fluid chemistry node. The result shows a 22% increase in the UDC rate with a corresponding increase in the expected utility for the production loss and environmental impact to 1.74266×10^7 USD and 1.00414×10^8 USD, respectively.

Further demonstration of the impact of a catastrophic failure on the economic losses is shown in Fig. 7.5. The analysis indicates that at a 100% likelihood of leak failure, the expected utility for the production loss/repair cost and the environmental impact, cleaning, and loss of reputation cost increases. This represents a 4.9 and

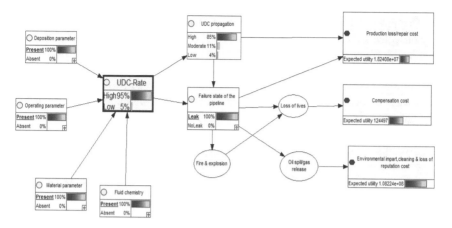

Fig. 7.5 Impact of catastrophic leak failure on the predicted cost-based consequences

8.3% increase in the economic losses due to loss in production and environmental impact, clean up, and loss of reputation, respectively. The expectation decision theory provides a cost-based estimation tool for the failure consequences analysis for the given scenarios. This provides a handful of information for operational planning and mitigation management of critical oil and gas assets against failure. The approach supported asset life cycle appraisal for a robust integrity management strategy.

7.4 Conclusions

The current work presents an intelligent cost-based tool based on BN and expected utility decision theory for failure consequence analysis in harsh offshore operations. The complexity of the failure-induced events (e.g., UDC) presents a stochastic degradation process that requires a dynamic probabilistic approach. The BN model shows the capacity to capture the multidimensionality in the failure influencing factors to predict the rate of degradation and health state of the asset. The consequences of failure were then assessed based on the failure state of the asset to establish a cost-based prediction using the expected utility decision theory. The integration of the expected utility analysis provides a cost-value to the impact of failure on the operations, environment, and humans. The following outline shows the key findings of the present study.

- The dynamic and updating capabilities of the model make it adaptive and able to explore complex multidimensional interactions
- The model structural learned the key UDC influential factors to predict the rate degradation of the asset. This can be dynamically updated as the condition of the asset changes to explore the likelihood of leak failure and its influence on the consequences.

- The proposed model provides a robust structure that captures the two-level consequence prediction given the state of the asset under UDC
- The integrated expected utility analysis provides a cost-value to the various consequences of failure, which are updated based on the amount of spill. This changes with the degree of interaction among the influential risk factors
- It was observed that at the UDC rate of 0.755, the economic losses predicted as consequences of failure give 1.74×10^7 USD and 9.99×10^7 USD for the loss of production and environmental impact cost, respectively.
- Given the complex mechanism of UDC and its consequences, the present model provides a useful cost-based assessment tool that could guide a well-informed integrity management strategy for oil and gas operations in harsh environments

References

1. Adumene, S., Adedigba, S., Khan, F., Zendehboudi, S.: An integrated dynamic failure assessment model for offshore components under microbiologically influenced corrosion. Ocean Eng. **218**, 108082 (2020). https://doi.org/10.1016/j.oceaneng.2020.108082
2. PHMSA: Data and Statistics (2017)
3. Liu, E., Lv, L., Yi, Y., Xie, P.: Research on the steady operation optimization model of natural gas pipeline considering the combined operation of air coolers and compressors. IEEE Access (2019). https://doi.org/10.1109/ACCESS.2019.2924515
4. Su, Z., Liu, E., Xu, Y., Xie, P., Shang, C., Zhu, Q.: Flow field and noise characteristics of manifold in natural gas transportation station. Oil Gas Sci. Technol. (2019). https://doi.org/10.2516/ogst/2019038
5. Yazdi, M., Khan, F., Abbassi, R., Quddus, N., Castaneda-Lopez, H.: A review of risk-based decision-making models for microbiologically influenced corrosion (MIC) in offshore pipelines. Reliab. Eng. Syst. Saf. **223**, 108474 (2022). https://doi.org/10.1016/j.ress.2022.108474
6. El-Sherik, A.M.: Trends in Oil and Gas Corrosion Research and Technologies Production and Transmission, 1st ed. Elsevier (2017)
7. Huang, S.N.J., Brown, B., Jiang, X., Kinsella, B.: Internal CO_2 Corrosion of Mild Steel Pipelines Under Inert Solid Deposits (2010)
8. Menendez, D.S.C.M., Jovancicevic, V., Ramachandran, S., Morton, M.: Assessment of corrosion under iron sulfide deposits and CO_2/H_2S conditions. Corros. J. **69**(2), 145–156 (2013)
9. Peacock, D.K., Grauman, J.S.: Crevice and under deposit corrosion resistance of titanium alloys in highly aggressive environments. Mater. Corros. Werkstoffe und Korrosion **49**(2), 61–68 (1998). https://doi.org/10.1002/(SICI)1521-4176(199802)49:2%3c61::AID-MACO61%3e3.0.CO;2-I
10. Sliem, M.H., et al.: Monitoring of under deposit corrosion for the oil and gas industry: a review. J. Pet. Sci. Eng. **204**, 108752 (2021). https://doi.org/10.1016/j.petrol.2021.108752
11. Gupta, J., Talukdar, M.K., Velusshami, S.K., Sharma, A., Makkar, S.: Premature failure of submarine well fluid lines: a case study. J. Fail. Anal. Prev. **21**(2), 363–369 (2021). https://doi.org/10.1007/s11668-020-01051-0
12. Adumene, S., Okwu, M., Yazdi, M., Afenyo, M., Islam, R., Orji, C.U., Obeng, F., Goerlandt, F.: Dynamic logistics disruption risk model for offshore supply vessel operations in Arctic waters. Marit. Transp. Res. **2**, 100039 (2021). https://doi.org/10.1016/j.martra.2021.100039
13. Daly, R., Shen, Q., Aitken, S.: Learning Bayesian networks: approaches and issues. Knowl. Eng. Rev. **26**(2), 99–157 (2011). https://doi.org/10.1017/S0269888910000251

14. Beretta, S., Castelli, M., Gonçalves, I., Henriques, R., Ramazzotti, D.: Learning the structure of Bayesian networks: a quantitative assessment of the effect of different algorithmic schemes. **2018**(1), (2017)
15. Chickering, D.M.: Learning equivalence classes of Bayesian-network structures. J. Mach. Learn. Res. **2**(3), 445–498 (2002). https://doi.org/10.1162/153244302760200696
16. Adumene, S., et al.: Dynamic logistics disruption risk model for offshore supply vessel operations in Arctic waters. Marit. Transp. Res. **2**, 100039 (2021). https://doi.org/10.1016/j.martra.2021.100039
17. Yazdi, M., Khan, F., Abbassi, R., Quddus, N.: Resilience assessment of a subsea pipeline using dynamic Bayesian network. J. Pipeline Sci. Eng. **2**, 100053 (2022). https://doi.org/10.1016/j.jpse.2022.100053
18. Yazdi, M., Khan, F., Abbassi, R.: Operational subsea pipeline assessment affected by multiple defects of microbiologically influenced corrosion. Process Saf. Environ. Prot. **158**, 159–171 (2021). https://doi.org/10.1016/j.psep.2021.11.032
19. Yazdi, M., Khan, F., Abbassi, R.: Microbiologically influenced corrosion (MIC) management using Bayesian inference. Ocean Eng. (2021). https://doi.org/10.1016/j.oceaneng.2021.108852
20. Yazdi, M., Khan, F., Abbassi, R.: A dynamic model for microbiologically influenced corrosion (MIC) integrity risk management of subsea pipelines. Ocean Eng. **269**, 113515 (2023). https://doi.org/10.1016/j.oceaneng.2022.113515
21. Adumene, S., Islam, R., Dick, I.F., Zarei, E., Inegiyemiema, M., Yang, M.: Influence-based consequence assessment of subsea pipeline failure under stochastic degradation. Energies **15**, 20 (2022). https://doi.org/10.3390/en15207460
22. Savage, L.J.: The Foundations of Statistics, 2nd ed. Dover Publications Inc., New York (1972)
23. Okoro, C., Ekun, O.A., Nwume, M.I., Lin, J.: Molecular analysis of microbial community structures in Nigerian oil production and processing facilities in order to access souring corrosion and methanogenesis. Corros. Sci. **103**, 242–254 (2016)
24. Adumene, S., Khan, F., Adedigba, S., Zendehboudi, S.: Offshore system safety and reliability considering microbial influenced multiple failure modes and their interdependencies. Reliab. Eng. Syst. Saf. **215**, 107862 (2021). https://doi.org/10.1016/j.ress.2021.107862
25. Li, H., Yazdi, M.: Advanced Decision-Making Methods and Applications in System Safety and Reliability Problems. Springer, Cham (2022)
26. Adumene, S., Nwoaha, T.C.: Dynamic cost-based integrity assessment of oil and gas pipeline suffering microbial induced stochastic degradation. J. Nat. Gas Sci. Eng. **96**, 104319 (2021). https://doi.org/10.1016/j.jngse.2021.104319
27. Heredia-Zavoni, E., Montes-Iturrizaga, R., Faber, M.H., Straub, D.: Risk assessment for structural design criteria of FPSO systems. Part II: Consequence models and applications to determination of target reliabilities. Mar. Struct. **28**(1), 50–66 (2012). https://doi.org/10.1016/j.marstruc.2012.05.001
28. Yazdi, M., Nedjati, A., Zarei, E., Abbassi, R.: A novel extension of DEMATEL approach for probabilistic safety analysis in process systems. Saf. Sci. **121**, 119–136 (2020). https://doi.org/10.1016/j.ssci.2019.09.006
29. Richardson, R., Brugnone, N.: Oil Spill Economics: Estimates of the Economic Damages of an Oil Spill in the Straits of Mackinac in Michigan (2018). Available: https://flowforwater.org/wp-content/uploads/2018/05/FLOW_Report_Line-5_Final-release-1.pdf
30. Daneshvar, S., Yazdi, M., Adesina, K.A.: Fuzzy smart failure modes and effects analysis to improve safety performance of system: case study of an aircraft landing system. Qual. Reliab. Eng. Int. (2020). https://doi.org/10.1002/qre.2607
31. Abbassi, R., Arzaghi, E., Yazdi, M., Aryai, V., Garaniya, V., Rahnamayiezekavat, P.: Risk-based and predictive maintenance planning of engineering infrastructure: existing quantitative techniques and future directions. Process Saf. Environ. Prot. **165**, 776–790 (2022). https://doi.org/10.1016/j.psep.2022.07.046

Chapter 8
A Sustainable Circular Economy in Energy Infrastructure: Application of Supercritical Water Gasification System

Abstract Integrating circular economy (CE) and energy infrastructure (EI) is considered a critical part of the sustainable development of societies. The present EIs have been challenged with different kinds of technical and economic aspects over the system lifecycle, including recourses misuse, pointless competition between end-user parties, and increasing decommissioned waste materials. Recently, decision-makers, policymakers, scholars, engineers, and practitioners have been interested in integrating CE and IEs. Remarkably, there is still a lack of knowledge regarding a solid connection between CE, IEs, and other contributing parameters. The main idea behind the CE is to make better products, materials, and system components utilization throughout their sequential lifecycles. In this chapter, the aim is to investigate sustainable CE in the design and development of EIs. Thus, the Risk Management (RM) of a supercritical water gasification system as a critical EI has been studied to identify the relevant implication in practice, gaps in the existing knowledge, and cutting-edge strategies to empower the CE in the lifecycle of EIs, such as recertification, refurbishment and remanufacturing, recycling, to decommission, landfill, and more. Finally, the present study provides a series of recommendations for research activities as well as innovative management practices to take industrial and governmental sectors.

Keywords Failure mode and effect analysis · Decision-making · Supercritical water gasification system

8.1 Introduction

Over the last few years, several practitioners and scholars have broadly discussed the integration subjects of circular economy (CE) and energy infrastructure (EI) [1]. However, this topic is commonly investigated individually without consideration of other aspects (e.g., risk management, system safety, reliability, and resilience assessment of critical energy infrastructures). Performing Risk Management (RM) is crucial because it is an essential task for CE and IE integration in practice throughout the lifecycle of the system. In addition, RM eliminates or reduces the adverse effects of waste materials (i.e., potential hazards) in the circular economy [2, 3]. Kirchherr

© The Author(s), under exclusive license to Springer Nature Switzerland AG 2023   119
H. Li et al., *Intelligent Reliability and Maintainability of Energy Infrastructure Assets*,
Studies in Systems, Decision and Control 473,
https://doi.org/10.1007/978-3-031-29962-9_8

et al. [4] carried out a comprehensive review of the definitions of CE. The present chapter considered the CE definition provided by Preston and Lehne [5]: "The basic idea of the CE is to shift from a system in which resources are extracted, turned into products and finally discarded towards one in which resources are maintained at their highest value possible". With increasing urbanization and worldwide population, the waste materials like sewage sludge are among one the most significant concerns in modernized societies, which enface individuals with several types of health and safety problems (e.g., hazardous and toxic elements: mercury, chromium, lead, etc.) [6] and impacts on public health and environmental ecosystem [7, 8].

The necessity of CE can also be highlighted as a better alternative to the existing linear "take–make–use–dispose economy" [9]. Besides, the CE significantly adds value to the system's sustainable development by minimizing waste and natural resource usage and minimizing the social, technical, and environmental aspects of the lifetime of the products and materials. In the existing literature, CE has brought a mixture of different ideas, including Zero-waste [10–12], resource efficiency [13, 14], cleaner production [15, 16], industrial ecology [17–19], and more [20–22]. It should be added that the CE aims to target four main strategies in the lifecycle of the system [23, 24]:

i.   Making economic resource flows to be narrow,
ii. Making the resource flows to be slow,
iii. Making resource flows be closed within material recycling, and
iv. Making resource flows be integrated into the biogeochemical cycles at the end of their usage.

This chapter aims to investigate sustainable CE in the design and development of EIs, by engaging it with RM of supercritical water gasification systems as a critical EI. According to this point, a systematic CE integration is required to overview the potential and applicable strategic management practices on supercritical water gasification systems.

In the next couple of sub-sections, an overview of recent selective published studies is carried out to highlight the most impactful contemporary subjects in integrating CE, IE, and RM.

8.1.1 Background: Circular Economy and Risk Management

To perform an overview analysis, in the Web of Science (WoS), the following search is considered: TOPIC, "Circular Economy" AND "Risk Management", by the end of December 2022. As can be seen from Fig. 8.1, the TreeMap chart of application areas demonstrated that CE and RM integration are prioritized in Environmental Sciences (20 documents), Management (19 documents), and Green Sustainable Science Technology (15 documents).

Figure 8.2 shows the number of publications and citations. In the last two years, 2021, and 2022, this integration area has received much more attention from

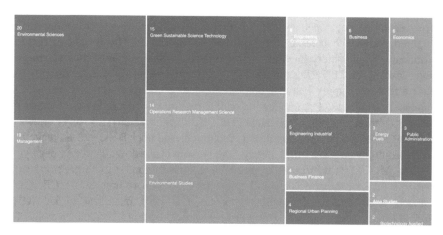

Fig. 8.1 The TreeMap chart of application areas constructed with CE, and RM integration studies (*Note* The areas on the chart are not strictly proportional to the values of each entry)

researchers. Therefore, it is predicted that the growth path will continue in the upcoming years.

In the existing literature, the role of carbon capture technologies in acquiring sustainable development and sustainable development goals was investigated [25]. In this study, authors attempted to identify several indicators in CE with carbon capture technologies to considerably help decision-makers in terms of RM improvement, data management enhancement, resource allocation and environmental performance improvement, and more items. According to the WoS, this study has received 25.5 Average citations per year (ACY). Another study proposes a legal framework [26] (has attracted 8.2 ACY) for chemical risk management and facilitating a CE.

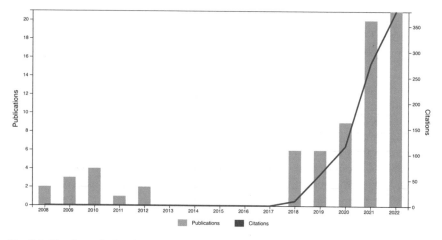

Fig. 8.2 The times cited and publications over time (CE, and RM integration studies)

The researchers developed an innovative scheme to describe how performing RM is required to reuse the material with hazardous substances. Finally, novel concepts for sustainability and safety integration are proposed, which support future CE transition. Yazdani et al. [27] (has attracted 20.0 ACY) investigated the flood risk drivers, their effects on supply chain agriculture, and their interconnection with the CE applicable strategies. The application of this research is underlying the context of real agricultural projects. Deshpande et al. [28] (has attracted 11.3 ACY) assessed different aspects (e.g., environmental, economic, and social) as well as recycling waste fishing gears based on Norway. The authors utilized multi-criteria decision-making (MCDM) to prioritize the end-of-life alternatives and their ability to manage waste plastics, sustainably. This study provides several opportunities for a CE establishment in the region of Scandinavia. Zhao et al. [29] (has attracted 3.4 ACY) studied safety vulnerability assessment for chemical enterprises' data envelopment analysis with fuzzy decision-making. Zhong et al. [30] (has attracted 3.0 ACY) explored emerging human resources with performing RM during Covid time across numerous industries. The study results indicated that Covid considerably impacts human resource management and needs significant theoretical and empirical consideration from scientists. In another study [31] (which has attracted 1.0 ACY), authors evaluated the impact of the occupancy variables within risk assessment in a low-energy. Finally, Rejeb et al. [32] studied the barriers to blockchain acceptance in the CE. The authors adopted an integrated approach constructed based on the fuzzy Delphi and best–worst methods to analyze and rank the identified 16 barriers. The study presents reliable insights for decision-makers and policymakers, which be utilized to optimize blockchain implementations in the CE.

Reviewing an integration of CE and RM shows that no attempts have been made to use risk assessment and management techniques such as Failure mode and effect analysis, fault tree analysis, Bayesian network, Bow-tie, risk matrix, and so more.

8.1.2   Background: Circular Economy and Energy Infrastructure

To perform an overview analysis, in the Web of Science (WoS), the following search is considered: TOPIC, "Circular Economy" AND "Energy Infrastructure", by the end of December 2022. As can be seen from Fig. 8.3, the TreeMap chart of application areas demonstrated that CE and IE integration are prioritized in Energy Fuels (5 documents), Environmental Sciences (4 documents), and Economics (2 documents).

Figure 8.4 displays the number of publications and citations. In the years 2020, and 2021, this integration area has received much more attention from researchers. Therefore, it seems that shortly, more attention will be required from the academia and industrial sectors.

In the existing literature, the link between modularization and CE is studied [33] (which has attracted 3.5 ACY) to identify the enabling factors and barriers for module

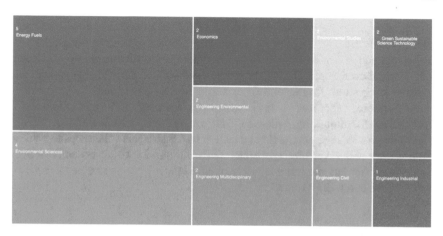

Fig. 8.3 The TreeMap chart of application areas constructed with CE, and EI integration studies (*Note* The areas on the chart are not strictly proportional to the values of each entry)

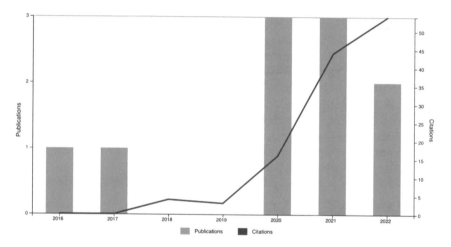

Fig. 8.4 The times cited and publications over time (CE, and EI integration studies)

reuse of components. The study results indicated that the stakeholders involved in planning, building, operating, and decommissioning EIs could get advantages of modular CE. Invernizzi et al. [34] (which has attracted 6.0 ACY) presented a piece of basic information about interdisciplinary thinking that is needed for decommissioning policy integration into CE concepts. To maximize the lifestyle value of EIs over time. The authors concluded that new research streams need to be recommended to promote much more sustainable management of EIs life. In this regard, the paper proposed several challenges and encouraged discussion on practical policy development for the existing and future decommissioning schemes. In another study,

Mignacca et al. [1] (which has attracted 8.33 ACY) reviewed the knowledge shortages between the modularisation CE and EIs. Following that, an investigation is carried out on the related demonstrations, including construction waste and closed-loop material cycle achievement. The study also discussed the available policies and provided suggestions for modular CE fostering in EIs for the future.

Furthermore, Moore et al. [35] (which has attracted 10.0 ACY) studied CE management for lithium-ion batteries by integrating three well-known MCDM tools. The study provides the potential contribution of CE strategies and helps to achieve the resilience and sustainability objectives. The lithium-ion batteries can be derived from waste materials and can offer a meaningful influence for post-disaster recovery through distributed EIs worldwide.

8.1.3 Background: Energy Infrastructure and Risk Management

To perform an overview analysis in the Web of Science (WoS), the following search is considered: TOPIC, "Risk Management" AND "Energy Infrastructure", by the end of December 2022. As can be seen from Fig. 8.5, the TreeMap chart of application areas demonstrated that RM and IE integration are prioritized in Energy Fuels (5 documents), Green Sustainable Science Technology (6 documents), and Environmental Science (4 documents).

Figure 8.6 presents the number of publications and citations. In 2021, this integration area received much more attention from researchers in publications and citations. Hence, it is predicted that more researchers and practitioners will work in this integration area in subsequent years.

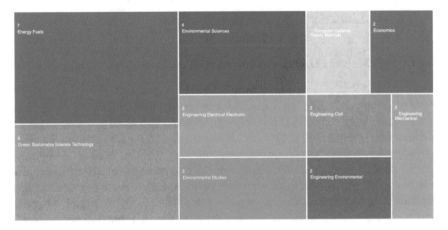

Fig. 8.5 The TreeMap chart of application areas constructed with EI, and RM integration studies (*Note* The areas on the chart are not strictly proportional to the values of each entry)

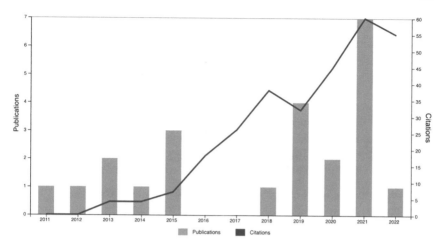

Fig. 8.6 The times cited and publications over time (RM, and EI integration studies)

An asset-level and future-based climate risk assessment approach is proposed in the existing literature [36]. In this research work, the authors developed a framework for the climate resilience pricing of an EI through the debt and equity investment values assessment. The execution of this approach indicates how to outline climate-related risks into financial-based ones. An application of study is also applied to the utility electricity generation unit within natural gas. In another study, a risk-based breakdown matrix is proposed to evaluate and assess the onshore wind farm project using Fuzzy case-based reasoning [37]. The outcome of the work highlighted that the introduced risk assessment approach over the conventional risk assessment techniques is much more applicable and practical for innovative kinds of construction projects with uncertain information. The "Horizontal Directional Drilling" risk assessment is conducted by developing an approach using a machine learning model [38]. In this research work, three different types of machine-learning tools ("logistic regression, random forests, and Artificial Neural Network") are utilized to estimate the risk of "Horizontal Directional Drilling" projects as well as identify the probability of undesired event occurrences. In another study [39], a climate risks overview of IEs in India is determined, and a series of recommendations are suggested, such as (a) "mandatory vulnerability assessment to future climate risks for energy infrastructures"; (b) "project and systemic risks in the vulnerability index"; (c) "adaptation funds for unmitigated climate risks"; (d) "continuous monitoring of climatic parameters and implementation of adaptation measures", and (e) "sustainability actions along energy infrastructures that enhance climate resilience and simultaneously deliver co-benefits to local agents". Hitzeroth and Megerle [40] proposed possible mechanisms to identify these especially high-risk social groups and discussed the assets and disadvantages of diverse management strategies to address the acceptable risks. This is crystal clear that the results consistently ought to be analyzed above the knowledge of the specific local and regional context, respectively. Finally, in an investigation-based study [41], a corresponding quantification method engages a hybrid tool with an integration of complicated network criteria with data analytics. To highlight the applicability of the proposed method, the City of Minneapolis bus transit network is assessed considering the possible overflow failures characterized by node, link, and route failure. In this study, the authors aim to provide a better understanding of system transitions and pre/post disruptions. In addition, the main significant key component is identified, and the network robustness is evaluated. At the end of the day, a viable systematic risk management strategy plan is suggested to empower the city's resilience through risk management.

The main contribution of the present chapter is triple as the following:

- An individual pairwise review of CE, IE, and RM has been carried out,
- Failure mode and effect analysis technique as a robust RM tool is conducted to improve system safety and reliability of the Supercritical water gasification system, and
- A crystal-clear response is provided to the question, *"what we know?"* about the connection and dependency among CE, EI, and RM.

The organization of this chapter is provided as the following. In the next Section, the application of the study by investigating and conducting the risk management of supercritical water gasification system is presented. Finally, in the last Section, the conclusion highlighting the remarks of the present work and recommending future directions are provided.

8.2 Discussion

In the wastewater treatment process, a large amount of sewage sludge is generated, referring to the semifluid waste with a range of solid substances (e.g., a rate of 0.25–12% rate depends on the treatment process) [42]. The sewage sludge carries heavy metals, microorganisms, and different organic compounds, resisting biological and natural decompositions. It is vital to consider that in case of improper sewage sludge treatment, includes significant sources of material pollutants and can harm the environment. In this regard, there is an essential for the wastewater treatment process that could eliminate the negative side effects on humans and the environment. In addition, in the process, we should ensure that the appropriate recovery process of macronutrients is found in substances compounds. An effective supercritical water gasification system can minimize the negative influence of sewage sludge on the environment, secure sustainable waste materials, and play a significant role in the circular economy. However, the supercritical water gasification system is risky because the process is operated with high temperature and pressure [43, 44]. Hence, it is essential to perform a risk management process (in this study, FMEA) to identify the potential hazards and provide a series of intervention actions to reduce the amount of risk to an acceptable level [45–47]. Table 8.1 presents the FMEA analysis of the supercritical water gasification system.

Table 8.1 FMEA analysis of supercritical water gasification system[a]

Tag	Failure modes	Severity	Occurrence	Detection	RPN
FM1	Incompatible fittings	4	7	2	56
FM2	Incompatible materials	5	5	5	125
FM3	Reactor	8	2	4	64
FM4	Design	3	2	5	30
FM5	Resistance	4	4	3	48
FM6	**Corrosion**	**5**	**6**	**4**	**120**
FM7	Plugging	3	3	6	54
FM8	**Heat exchanger**	**5**	**6**	**8**	**240**
FM9	Coolant temperature	4	5	5	100
FM10	Feedstock	2	4	5	40
FM11	Precipitation/not homogenous flow	5	4	2	40
FM12	Composition	5	2	3	30
FM13	Operator	4	5	5	100
FM14	**Improper maintenance**	**4**	**6**	**7**	**168**
FM15	Malfunction/fatigue	4	5	4	80
FM16	Landsman	3	4	6	72
FM17	Operation conditions	5	6	4	120
FM18	Pressure	5	6	3	90
FM19	Temperature	4	4	1	16
FM20	Solid matter content	4	3	5	60
FM21	Catalyst	4	5	4	80
FM22	Control panel/electricity short circuit	5	7	3	105
FM23	Breakdown	2	3	5	30
FM24	Environment/odour/noise	2	5	5	50
FM25	Gas–liquid separation	3	4	4	48

[a] The RPN is computed based on expert judgment, which has a relevant background and knowledge in CE, RM, and IE

According to the obtained results, one can see that the highest rank of failure modes is FM8, FM14, and FM6. Thereafter, a series of recommendations as intervention actions need to be presented, such as enhancing the periodical maintenance actions for all three failure modes. The point that should be considered is that the intervention actions must be under the context of CE as much as possible. In practice, there might be many management practices that are not CE-friendly and can harm the whole of the system cycle. Hence, this should be avoided from a set of suggested intervention actions when planning to have an integration of CE, IE, and RM.

Keeping the FMEA, as mentioned earlier in mind, the CE enfaces several system challenges and changes. However, conceptually, this is all about waste and resource management. Some of the challenges/opportunities of CE in critical IEs can be highlighted as the following:

- Fossil fuels: it is essential to move forward to use non-renewable energy since fossil-based one causes considerable global warming [48, 49]. Thus, reducing greenhouse gas emissions in the system's lifecycle is a severe challenge [50].
- Resource competition: it is an increase across the EI sectors, especially in batteries, large-scale energy storage, and grid development [51, 52]. This challenge enfaces the industrial and governmental sectors to break down into innovative global markets, such as biomass gasification [53, 54], hydrogen energy [55–57], offshore wind farms [58–61], IE material degradation [62, 63], and so on.
- Resource exploitation: Over the past decades, the demand for renewable energy has widely increased, enhancing the sustainability challenges as well [52]. In contrast, there are some opportunities in this area to develop some more relevant businesses, like novel and reliable start-ups
- Energy usage: As a matter of fact, engaging in renewable energy has increased nowadays; however, in most countries, energy usage is not feasible since it has to decline following the energy hierarchy [64].
- Lifetime extension and Durability: This is an essential part of resource productivity, which requires have considerable understanding of the system components and the whole part of EIs in their lifecycle, operation, and maintenance [65, 66]. Having reliable insights into material products and flows is essential for decision-making, policy developments, and investment into the CE [65].
- Material innovation: This part is also important and can positively impact sustainable CE, such as improving component durability, extending the production life of the system, enhancing the system resilience, substituting fossil-based fuels, easily recycling materials and compounds, and having a proper alternative for raw materials usage.
- Sustainable decommissioning: In this example, the decomposition experience is restricted since it is commonly taken into account as a reconsideration [67], and can easily design for decommissioning proactively. In the upcoming years, the need for more applicable regulations and policies increases the uncertainty in the decommissioning process and leads to improper plans. In addition, the lack of proactive planning in practice can impact the cost of decommissioning; in the case example, the costs could be five to ten times higher than previously expected [67].
- End-of-use management: There is a huge growth concern about resource access, as well as social/environmental impacts for end-of-use management purposes. However, concerns about end-of-use management are prevalent [68]. Long-running CE heightens these challenges regarding recycling end-user materials [69]. Thus, the supply chains for supercritical water gasification system decommissioning, related IEs, and *"end-of-use management"* are yet to be devised.

In this regard, the objective is presented in Fig. 8.7.

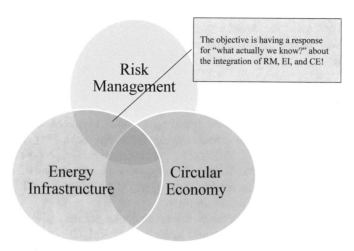

The objective is having a response for "what actually we know?" about the integration of RM, EI, and CE!

Fig. 8.7 The objectives of what actually we know about the integration of RM, EI, and CE

In the following, a couple of remarks are highlighted to have a well-implemented and appropriate integration of RM, CE, and IE in practice.

• Having a reliable closed-loop material cycling strategy

The existing literature highlights that the disassembly/deconstruction design is vital for a closed-loop material cycling strategy [68]. It is recognized that performing RM significantly improves the deconstruction boosting closed-loop material cycling system. In addition, it is required to have straightforward deconstruction instructions and procedures. However, there needs to be more information and details regarding the practical implication of RM in the closed-loop material cycling strategy.

• The need for lifecycle energy reduction

One of the main strategies for lifecycle energy reduction is prefabrication. In a study conducted by Aye et al. [70], lifecycle energy reduction is examined in three different construction forms (i.e., concrete-based, steel-based prefabricated, and timber-based). In this investigation, it is indicated that the embodied energy in steel-based prefabricated is 50% greater than the two other types of construction forms. Therefore, reusing the steel-based structure in a newly constructed building can save more than 80% of the same used energy.

• The necessity of construction demolition waste reduction

Some evidence shows that the prefabrication structure has enough potential and capability to reduce construction demolishing waste [71]. However, there needs to be more information on how it could impact the system, either environmentally or economically. A study [72] evaluated the impact of prefabrication on construction demolishing waste in China. The investigation revealed that using prefabrication considering all significant potential hazards can strongly impact the prefabrication promotion utilization as well as construction demolition waste underlying the economical aspects (e.g., tax income).

- Experiencing the direct recycling process

The cost of the recovery process, logistics, and environmental load could be reduced when the recycling process is used directly (i.e., there is no need for components' disassembly) [73]. In a study carried out by Fukushige et al. [74], a framework is developed according to the different lifecycle scenarios. The proposed approach considers the components characterized as adequate for the equivalent lifestyle alternatives by ignoring the disassembly and evaluating the integration of CE and IE underlying the context of resource efficiency.

Considering the recommendation mentioned above, we may now have a better vision of integrating the three essential parameters of RM, CE, and IE, and have a couple of ideas and responses in mind for "*what we know*?" about the integration of RM, EI, and CE!

8.3   Conclusion

In this chapter, we attempted to provide an apparent response to the question, "*what do we know*?" about the connection and dependency among Circular Economy, Energy Infrastructure, and Risk Management. Although there is considerable attention from decision-makers, scholars, and policymakers in both industrial and governmental sectors, there are still rooms to make further efforts on such interdependency investigation. In the state of the arts, there are a limited number of published works highlighting different aspects of connections in (CE and IE), and (IE and RM); however, no publications bring the idea of integration of CE, IE, and RM together.

The nationwide capacities concerning the different CE strategies impact the lifecycle feasibility. In such cases, when the capability for maintainability and operability is low, thereafter, relying on maintainability and repair management practices is not suggested to be executed. In contrast, a heavy concentration is needed on the maintainability and repair management practices to enhance the lifetime extension when the disassembly, recycling, and decommissioning capabilities are weak and underdeveloped.

In addition, the integration of CE and IE within supercritical water gasification systems requires supportive decision-making methods for the entire lifecycle of the system, determining which type of strategy is the best fit and feasible and under which circumstances. Based on the various legal condition worldwide, supportive decision-making methods have to cooperate with both industrial and governmental sectors.

In addition, the uptake and CE development of supercritical water gasification systems needs to contain a more profound and much more comprehensive baseline assessment using a literature review and industrial practitioners and researchers working on different kinds of CE strategies. This includes but is not limited to exchanging decision-makers' knowledge on CE, IE, and RM under the supercritical water gasification system. Such these types of discussions are also informative and can clarify the enablers/barriers for CE in supercritical water gasification systems.

Finally, the discussed CE framework in this study could improve the management sustainability of the IEs. However, as the paper explains with a critical IE example, it is only possible to address some sustainability challenges. Integrating the IEs into discussions between CE and RM will need a more comprehensive engagement exceeding the material characteristics of developing IEs.

References

1. Mignacca, B., Locatelli, G., Velenturf, A.: Modularisation as enabler of circular economy in energy infrastructure. Energy Policy **139**, 111371 (2020). https://doi.org/10.1016/j.enpol.2020.111371
2. Kazancoglu, Y., Ozkan-Ozen, Y.D., Mangla, S.K., Ram, M.: Risk assessment for sustainability in e-waste recycling in circular economy. Clean Technol. Environ. Policy **24**, 1145–1157 (2022). https://doi.org/10.1007/s10098-020-01901-3
3. Nedjati, A., Yazdi, M., Abbassi, R.: A Sustainable Perspective of Optimal Site Selection of Giant Air—Purifiers in Large Metropolitan Areas. Springer, Netherlands (2021). https://doi.org/10.1007/s10668-021-01807-0
4. Kirchherr, J., Reike, D., Hekkert, M.: Conceptualizing the circular economy: an analysis of 114 definitions, Resour. Conserv. Recycl. **127**, 221–232 (2017). https://doi.org/10.1016/j.resconrec.2017.09.005
5. Preston, F., Lehne, J.: A Wider Circle? The Circular Economy in Developing Countries, Chatham House. (n.d.). https://policycommons.net/artifacts/1423261/a-wider-circle-the-circular-economy-in-developing-countries/2037525/. 24 Dec 2022. CID: 20.500.12592/c5sgwh
6. Zeng, X., Mathews, J.A., Li, J.: Urban mining of e-waste is becoming more cost-effective than virgin mining. Environ. Sci. Technol. **52**, 4835–4841 (2018). https://doi.org/10.1021/acs.est.7b04909
7. Liu, Q., Shi, S.J., Du, L.Q., Wang, Y., Cao, J., Xu, C., Fan, F.Y., Giesy, J.P., Hecker, M.: Environmental and health challenges of the global growth of electronic waste. Environ. Sci. Pollut. Res. **19**, 2460–2462 (2012). https://doi.org/10.1007/s11356-012-0923-z
8. Awasthi, A.K., Zeng, X., Li, J.: Relationship between e-waste recycling and human health risk in India: a critical review. Environ. Sci. Pollut. Res. **23**, 11509–11532 (2016). https://doi.org/10.1007/s11356-016-6085-7
9. Velenturf, A.P.M.: A framework and baseline for the integration of a sustainable circular economy in offshore wind. Energies. **14**, 5540 (2021). https://doi.org/10.3390/en14175540
10. Chong, M.N., Jin, B., Chow, C.W.K., Saint, C.: Recent developments in photocatalytic water treatment technology: a review. Water Res. **44**, 2997–3027 (2010). https://doi.org/10.1016/j.watres.2010.02.039
11. Spooren, J., Binnemans, K., Björkmalm, J., Breemersch, K., Dams, Y., Folens, K., González-Moya, M., Horckmans, L., Komnitsas, K., Kurylak, W., Lopez, M., Mäkinen, J., Onisei, S., Oorts, K., Peys, A., Pietek, G., Pontikes, Y., Snellings, R., Tripiana, M., Varia, J., Willquist, K., Yurramendi, L., Kinnunen, P.: Near-zero-waste processing of low-grade, complex primary ores and secondary raw materials in Europe: technology development trends. Resour. Conserv. Recycl. **160**, 104919 (2020). https://doi.org/10.1016/j.resconrec.2020.104919
12. Binnemans, K., Jones, P.T., Blanpain, B., Van Gerven, T., Pontikes, Y.: Towards zero-waste valorisation of rare-earth-containing industrial process residues: a critical review. J. Clean. Prod. **99**, 17–38 (2015). https://doi.org/10.1016/j.jclepro.2015.02.089

13. Duflou, J.R., Sutherland, J.W., Dornfeld, D., Herrmann, C., Jeswiet, J., Kara, S., Hauschild, M., Kellens, K.: Towards energy and resource efficient manufacturing: a processes and systems approach. CIRP Ann. **61**, 587–609 (2012). https://doi.org/10.1016/j.cirp.2012.05.002
14. Allwood, J.M.: Unrealistic techno-optimism is holding back progress on resource efficiency. Nat. Mater. **17**, 1050–1051 (2018). https://doi.org/10.1038/s41563-018-0229-8
15. Almeida, C.M.V.B., Agostinho, F., Huisingh, D., Giannetti, B.F.: Cleaner production towards a sustainable transition. J. Clean. Prod. **142**, 1–7 (2017). https://doi.org/10.1016/j.jclepro.2016.10.094
16. Ghisellini, P., Cialani, C., Ulgiati, S.: A review on circular economy: the expected transition to a balanced interplay of environmental and economic systems. J. Clean. Prod. **114**, 11–32 (2016). https://doi.org/10.1016/j.jclepro.2015.09.007
17. Wallner, H.P., Narodoslawsky, M.: The concept of sustainable islands: cleaner production, industrial ecology and the network paradigm as preconditions for regional sustainable development. J. Clean. Prod. **2**, 167–171 (1994). https://doi.org/10.1016/0959-6526(94)900 39-6
18. Barnard, F.: Education for management conceived as a study of industrial ecology. Vocat. Asp. Educ. **15**, 22–26 (1963). https://doi.org/10.1080/03057876380000041
19. Jelinski, L.W., Graedel, T.E., Laudise, R.A., McCall, D.W., Patel, C.K.: Industrial ecology: concepts and approaches. Proc. Natl. Acad. Sci. **89**, 793–797 (1992). https://doi.org/10.1073/pnas.89.3.793
20. Suárez-Eiroa, B., Fernández, E., Méndez-Martínez, G., Soto-Oñate, D.: Operational principles of circular economy for sustainable development: linking theory and practice. J. Clean. Prod. **214**, 952–961 (2019). https://doi.org/10.1016/j.jclepro.2018.12.271
21. Korhonen, J., Honkasalo, A., Seppälä, J.: Circular economy: the concept and its limitations. Ecol. Econ. **143**, 37–46 (2018). https://doi.org/10.1016/j.ecolecon.2017.06.041
22. Geissdoerfer, M., Savaget, P., Bocken, N.M.P., Hultink, E.J.: The circular economy—a new sustainability paradigm? J. Clean. Prod. **143**, 757–768 (2017). https://doi.org/10.1016/j.jclepro.2016.12.048
23. Velenturf, A.P.M., Archer, S.A., Gomes, H.I., Christgen, B., Lag-Brotons, A.J., Purnell, P.: Circular economy and the matter of integrated resources. Sci. Total Environ. **689**, 963–969 (2019). https://doi.org/10.1016/j.scitotenv.2019.06.449
24. Bocken, N.M.P., de Pauw, I., Bakker, C., van der Grinten, B.: Product design and business model strategies for a circular economy. J. Ind. Prod. Eng. **33**, 308–320 (2016). https://doi.org/10.1080/21681015.2016.1172124
25. Olabi, A.G., Obaideen, K., Elsaid, K., Wilberforce, T., Sayed, E.T., Maghrabie, H.M., Abdelkareem, M.A.: Assessment of the pre-combustion carbon capture contribution into sustainable development goals SDGs using novel indicators. Renew. Sustain. Energy Rev. **153**, 111710 (2022). https://doi.org/10.1016/j.rser.2021.111710
26. Bodar, C., Spijker, J., Lijzen, J., Waaijers-van der Loop, S., Luit, R., Heugens, E., Janssen, M., Wassenaar, P., Traas, T.: Risk management of hazardous substances in a circular economy. J. Environ. Manage. **212**, 108–114 (2018). https://doi.org/10.1016/j.jenvman.2018.02.014
27. Yazdani, M., Gonzalez, E.D.R.S., Chatterjee, P.: A multi-criteria decision-making framework for agriculture supply chain risk management under a circular economy context. Manag. Decis. **59**, 1801–1826 (2021). https://doi.org/10.1108/MD-10-2018-1088
28. Deshpande, P.C., Skaar, C., Brattebø, H., Fet, A.M.: Multi-criteria decision analysis (MCDA) method for assessing the sustainability of end-of-life alternatives for waste plastics: a case study of Norway. Sci. Total Environ. **719**, 137353 (2020). https://doi.org/10.1016/j.scitotenv.2020.137353
29. Zhao, R., Liu, S., Liu, Y., Zhang, L., Li, Y.: A safety vulnerability assessment for chemical enterprises: a hybrid of a data envelopment analysis and fuzzy decision-making. J. Loss Prev. Process Ind. **56**, 95–103 (2018). https://doi.org/10.1016/j.jlp.2018.08.018
30. Zhong, Y., Li, Y., Ding, J., Liao, Y.: Risk management: exploring emerging human resource issues during the COVID-19 pandemic. J. Risk Financ. Manag. **14**, 228 (2021). https://doi.org/10.3390/jrfm14050228

31. Curado, A., Silva, J.P., Lopes, S.I.: Radon risk assessment in a low-energy consumption school building: a dosimetric approach for effective risk management. Energy Rep. **6**, 897–902 (2020). https://doi.org/10.1016/j.egyr.2019.11.155

32. Rejeb, A., Rejeb, K., Keogh, J.G., Zailani, S.: Barriers to blockchain adoption in the circular economy: a fuzzy Delphi and best-worst approach. Sustainability **14**, 3611 (2022). https://doi.org/10.3390/su14063611

33. Benito, M., Giorgio, L.: Modular circular economy in energy infrastructure projects: enabling factors and barriers. J. Manag. Eng. **37**, 4021053 (2021). https://doi.org/10.1061/(ASCE)ME.1943-5479.0000949

34. Invernizzi, D.C., Locatelli, G., Velenturf, A., Love, P.E.D., Purnell, P., Brookes, N.J.: Developing policies for the end-of-life of energy infrastructure: coming to terms with the challenges of decommissioning. Energy Policy **144**, 111677 (2020). https://doi.org/10.1016/j.enpol.2020.111677

35. Moore, E.A., Russell, J.D., Babbitt, C.W., Tomaszewski, B., Clark, S.S.: Spatial modeling of a second-use strategy for electric vehicle batteries to improve disaster resilience and circular economy. Resour. Conserv. Recycl. **160**, 104889 (2020). https://doi.org/10.1016/j.resconrec.2020.104889

36. In, S.Y., Weyant, J.P., Manav, B.: Pricing climate-related risks of energy investments. Renew. Sustain. Energy Rev. **154**, 111881 (2022). https://doi.org/10.1016/j.rser.2021.111881

37. Somi, S., Gerami Seresht, N., Fayek, A.R.: Developing a risk breakdown matrix for onshore wind farm projects using fuzzy case-based reasoning. J. Clean. Prod. **311**, 127572 (2021). https://doi.org/10.1016/j.jclepro.2021.127572

38. Krechowicz, M., Krechowicz, A.: Risk assessment in energy infrastructure installations by horizontal directional drilling using machine learning. Energies **14**, 289 (2021). https://doi.org/10.3390/en14020289

39. Garg, A., Naswa, P., Shukla, P.R.: Energy infrastructure in India: profile and risks under climate change. Energy Policy **81**, 226–238 (2015). https://doi.org/10.1016/j.enpol.2014.12.007

40. Hitzeroth, M., Megerle, A.: Renewable energy projects: acceptance risks and their management. Renew. Sustain. Energy Rev. **27**, 576–584 (2013). https://doi.org/10.1016/j.rser.2013.07.022

41. Rasha, H., Ahmed, Y., Mohamed, E., Wael, E.-D.: Robustness quantification of transit infrastructure under systemic risks: a hybrid network-analytics approach for resilience planning. J. Transp. Eng. Part A Syst. **148**, 4022089 (2022). https://doi.org/10.1061/JTEPBS.0000705

42. Adar, E., İnce, M., Karatop, B., Bilgili, M.S.: The risk analysis by failure mode and effect analysis (FMEA) and fuzzy-FMEA of supercritical water gasification system used in the sewage sludge treatment. J. Environ. Chem. Eng. **5**, 1261–1268 (2017). https://doi.org/10.1016/j.jece.2017.02.006

43. Nie, R., Tian, Z., Wang, X., Wang, J., Wang, T.: Risk evaluation by FMEA of supercritical water gasification system using multi-granular linguistic distribution assessment. Knowledge-Based Syst. **162**, 185–201 (2018). https://doi.org/10.1016/j.knosys.2018.05.030

44. Yazdi, M., Nedjati, A., Zarei, E., Abbassi, R.: A reliable risk analysis approach using an extension of best-worst method based on democratic-autocratic decision-making style. J. Clean. Prod. **256**, 120418 (2020). https://doi.org/10.1016/j.jclepro.2020.120418

45. Li, H., Yazdi, M.: A holistic question: is it correct that decision-makers neglect the probability in the risk assessment method? In: Li, H., Yazdi, M. (eds.) Advanced Decision-Making Methods and Applications in System Safety and Reliability Problems: Approaches, Case Studies, Multi-criteria Decision Making, pp. 185–189. Springer International Publishing, Cham (2022). https://doi.org/10.1007/978-3-031-07430-1_10

46. Li, H., Yazdi, M.: Dynamic decision-making trial and evaluation laboratory (DEMATEL): improving safety management system. In: Li, H., Yazdi, M. (eds.), Advanced Decision-Making Methods and Applications in System Safety and Reliability Problems: Approaches, Case Studies, Multi-criteria Decision-Making, pp. 1–14. Springer International Publishing, Cham (2022). https://doi.org/10.1007/978-3-031-07430-1_1

47. Li, H., Yazdi, M.: What are the critical well-drilling blowouts barriers? A progressive DEMATEL-game theory. In: Li, H., Yazdi, M. (eds.) Advanced Decision-Making Methods

and Applications in System Safety and Reliability Problems: Approaches, Case Studies, Multi-criteria Decision-Making, Multi-objective, pp. 29–46. Springer International Publishing, Cham (2022). https://doi.org/10.1007/978-3-031-07430-1_3

48. Stamford, L., Azapagic, A.: Life cycle sustainability assessment of UK electricity scenarios to 2070. Energy Sustain. Dev. **23**, 194–211 (2014). https://doi.org/10.1016/j.esd.2014.09.008
49. Amran, Y.H.A., Amran, Y.H.M., Alyousef, R., Alabduljabbar, H.: Renewable and sustainable energy production in Saudi Arabia according to Saudi Vision 2030; Current status and future prospects. J. Clean. Prod. **247**, 119602 (2020). https://doi.org/10.1016/j.jclepro.2019.119602
50. Widger, P., Haddad, A.: Evaluation of SF6 leakage from gas insulated equipment on electricity networks in Great Britain. Energies **11**, 2037 (2018). https://doi.org/10.3390/en11082037
51. Soukissian, T.H., Denaxa, D., Karathanasi, F., Prospathopoulos, A., Sarantakos, K., Iona, A., Georgantas, K., Mavrakos, S.: Marine renewable energy in the Mediterranean Sea: status and perspectives. Energies **10**, 1512 (2017). https://doi.org/10.3390/en10101512
52. Roelich, K., Dawson, D.A., Purnell, P., Knoeri, C., Revell, R., Busch, J., Steinberger, J.K.: Assessing the dynamic material criticality of infrastructure transitions: a case of low carbon electricity. Appl. Energy. **123**, 378–386 (2014). https://doi.org/10.1016/j.apenergy.2014.01.052
53. Guoxin, H., Hao, H.: Hydrogen rich fuel gas production by gasification of wet biomass using a CO_2 sorbent. Biomass Bioenergy **33**, 899–906 (2009)
54. Levin, D.B., Chahine, R.: Challenges for renewable hydrogen production from biomass. Int. J. Hydrogen Energy. **35**, 4962–4969 (2010). https://doi.org/10.1016/j.ijhydene.2009.08.067
55. Zarei, E., Khan, F., Yazdi, M.: A dynamic risk model to analyze hydrogen infrastructure. Int. J. Hydrogen Energy. **46**, 4626–4643 (2021). https://doi.org/10.1016/j.ijhydene.2020.10.191
56. Yazdi, M.: The application of bow-tie method in hydrogen sulfide risk management using layer of protection analysis (LOPA). J. Fail. Anal. Prev. **17**, 291–303 (2017). https://doi.org/10.1007/s11668-017-0247-x
57. Li, H., Yazdi, M.: Integration of the Bayesian network approach and interval type-2 fuzzy sets for developing sustainable hydrogen storage technology in large metropolitan areas. In: Li, H., Yazdi, M. (eds.) Advanced Decision-Making Methods and Applications in System Safety and Reliability Problem, pp. 69–85. Springer International Publishing, Cham (2022). https://doi.org/10.1007/978-3-031-07430-1_5
58. Paul, S., Nath, A.P., Rather, Z.H.: A multi-objective planning framework for coordinated generation from offshore wind farm and battery energy storage system. IEEE Trans. Sustain. Energy. **11**, 2087–2097 (2020). https://doi.org/10.1109/TSTE.2019.2950310
59. Zhong, S., Pantelous, A.A., Beer, M., Zhou, J.: Constrained non-linear multi-objective optimisation of preventive maintenance scheduling for offshore wind farms. Mech. Syst. Signal Process. **104**, 347–369 (2018). https://doi.org/10.1016/j.ymssp.2017.10.035
60. Yazdi, M., Nedjati, A., Zarei, E., Abbassi, R.: Application of multi-criteria decision-making tools for a site analysis of offshore wind turbines (Chapter 6). In: Asadnia, M., Razmjou, A., et al. (eds.) Cognitive Data Science in Sustainable Computing, pp. 109–127. Academic Press, Cambridge. https://doi.org/10.1016/B978-0-323-90508-4.00008-3
61. Salvador, C.B., Arzaghi, E., Yazdi, M., Jahromi, H.A.F., Abbassi, R.: A multi-criteria decision-making framework for site selection of offshore wind farms in Australia. Ocean Coast. Manag. **224**, 106196 (2022). https://doi.org/10.1016/j.ocecoaman.2022.106196
62. Li, H., Yazdi, M., Huang, H.-Z., Huang, C.-G., Peng, W., Nedjati, A., Adesina, K.A.: A fuzzy rough copula Bayesian network model for solving complex hospital service quality assessment. Complex Intell. Syst. (2023). https://doi.org/10.1007/s40747-023-01002-w
63. Yazdi, M., Khan, F., Abbassi, R., Quddus, N.: Resilience assessment of a subsea pipeline using dynamic Bayesian network. J. Pipeline Sci. Eng. **2**, 100053 (2022). https://doi.org/10.1016/j.jpse.2022.100053
64. IEA: World Energy Outlook. Paris (2020)
65. Purnell, P., Velenturf, A., Jensen, P., Cliffe, N., Jopson, J.: Developing Technology, Approaches and Business Models for Decommissioning of Low-Carbon Infrastructure: E4 LCID, Low Carbon Infrastructure. Decommissioning Work (2018). https://www.gov.uk/government/publications/nuclear-provision-explaining-the-cost-of-cleaning-up-britains-nuclear-legacy/nuclear-provision-explaining-the

66. Borthwick, A.G.L.: Marine renewable energy seascape. Engineering **2**, 69–78 (2016). https://doi.org/10.1016/J.ENG.2016.01.011
67. Topham, E., McMillan, D.: Sustainable decommissioning of an offshore wind farm. Renew. Energy. **102**, 470–480 (2017). https://doi.org/10.1016/j.renene.2016.10.066
68. Jensen, P.D., Purnell, P., Velenturf, A.P.M.: Highlighting the need to embed circular economy in low carbon infrastructure decommissioning: the case of offshore wind. Sustain. Prod. Consum. **24**, 266–280 (2020). https://doi.org/10.1016/j.spc.2020.07.012
69. Lichtenegger, G., Rentizelas, A.A., Trivyza, N., Siegl, S.: Offshore and onshore wind turbine blade waste material forecast at a regional level in Europe until 2050. Waste Manag. **106**, 120–131 (2020). https://doi.org/10.1016/j.wasman.2020.03.018
70. Aye, L., Ngo, T., Crawford, R.H., Gammampila, R., Mendis, P.: Life cycle greenhouse gas emissions and energy analysis of prefabricated reusable building modules. Energy Build. **47**, 159–168 (2012). https://doi.org/10.1016/j.enbuild.2011.11.049
71. Arslan, H., Coşgun, N., Salgın, B.: Construction and demolition waste management in Turkey. In: Rebellon, L.F.M. (ed.) Waste Management—An Integrated Vision. IntechOpen, Rijeka (2012). https://doi.org/10.5772/46110
72. Li, Z., Shen, G.Q., Alshawi, M.: Measuring the impact of prefabrication on construction waste reduction: an empirical study in China. Resour. Conserv. Recycl. **91**, 27–39 (2014). https://doi.org/10.1016/j.resconrec.2014.07.013
73. Umeda, Y., Fukushige, S., Tonoike, K.: Evaluation of scenario-based modularization for lifecycle design. CIRP Ann. **58**, 1–4 (2009). https://doi.org/10.1016/j.cirp.2009.03.083
74. Fukushige, S., Tonoike, K., Inoue, Y., Umeda, Y.: D23 product modularization and evaluation based on lifecycle scenarios (life cycle engineering and environmentally conscious manufacturing). Proc. Int. Conf. Lead. Edge Manuf. 21st Century LEM21 **5**, 511–516 (2009). https://doi.org/10.1299/jsmelem.2009.5.511

Chapter 9
Attention Towards Energy Infrastructures: Challenges and Solutions

Abstract The energy infrastructure sector faces numerous challenges, including integrating renewable energy, digitizing energy systems, energy storage, microgrids and community energy initiatives, energy market design, environmental sustainability, and cybersecurity. These challenges require ongoing research and development to optimize the design and operation of energy infrastructure, ensure cost-effectiveness and efficiency, and promote a sustainable energy future. In this chapter, considering field research, interviewing with the experts, to the best of our literature review, and with the help and utilization of AI tools (specially Openai), we noticed a series of energy infrastructure challenges and how to deal with them. As a result, future studies in the energy infrastructure sector should focus on grid optimization, energy demand management, decentralized energy systems, financing and investment, energy transition and low-carbon development, rural electrification, energy policy and governance, and public awareness and engagement. These studies will provide valuable insights and guidance for the continued development of a robust and sustainable energy infrastructure.

Keywords Digital technologies · Climate change · Grid optimization · Demand management · Cybersecurity

9.1 Introduction

Attention towards energy infrastructure is crucial in both developed and developing countries, as access to reliable and sustainable energy sources is fundamental for socio-economic growth and development. Energy infrastructure is the backbone of a country's economy, providing the power required for industry, commerce, transportation, and households [1–5]. In developed countries, attention to energy infrastructure is essential for maintaining the stability and reliability of energy systems, as well as for addressing environmental and sustainability concerns [6, 7].

In developing countries, attention towards energy infrastructure is even more critical, as access to energy is often limited, and energy systems often need to be revised or updated. Energy infrastructure development in these countries can help to alleviate poverty, drive economic growth, and improve the standard of living for people

[8–10]. For example, access to electricity can provide opportunities for education, healthcare, and job creation, which can have a significant impact on reducing poverty. Furthermore, the energy sector is a key driver of greenhouse gas emissions and plays a substantial role in climate change. Developing countries are particularly vulnerable to the impacts of climate change, and investment in renewable energy and low-carbon energy infrastructure can help to mitigate these impacts and promote sustainable development. Additionally, engagement towards energy infrastructure is also important for energy security. Energy security refers to the availability and reliability of energy sources and the protection of energy supply chains. Developed countries often rely on energy imports, making them vulnerable to fluctuations in energy prices and disruptions to supply [11, 12].

In contrast, developing countries may suffer from energy poverty, with many people lacking access to essential energy services, such as lighting, heating, and cooking. Improving energy infrastructure in these countries can enhance energy security by reducing dependence on energy imports and increasing access to energy services. Investment in energy infrastructure also has the potential to create new economic opportunities and jobs. For example, the growth of renewable energy technologies [13], such as wind and solar power, has created new job opportunities in manufacturing, installation, and maintenance areas. In developing countries, investment in energy infrastructure can help to diversify the economy, create new jobs, and stimulate economic growth. Another reason for the importance of energy infrastructure is that it is closely linked to access to essential services, such as healthcare and education. Access to electricity is necessary for powering hospitals, schools, and other critical services. In developing countries, many people lack access to basic services due to a lack of energy infrastructure, and investment in this area can help improve people's quality of life.

Thus, developing awareness of energy infrastructure underlying the concept of advanced decision-making approaches [14–16] with consideration of uncertainty handling [17, 18] is essential in both developed and developing countries. Access to reliable and sustainable energy is crucial for socioeconomic growth and development, and the energy sector is a key driver of greenhouse gas emissions and climate change [19]. In this regard, the main contribution of the present chapter is to identify the main challenges that energy infrastructures have faced and provide a series of treatments to deal with them.

In Sect. 9.2, the primary identified challenges of energy infrastructures are presented. Section 9.3 provides a series of solutions as a treatment approach. Section 9.4 discusses the role of artificial intelligence in empowering energy infrastructures. The conclusion and future remarks are presented in Sect. 9.5.

9.2   The Identified Challenges of Energy Infrastructures

In this section, the main challenges of energy infrastructures in both developed and developing countries are identified considering field research, interviewing with the experts, to the best of our literature review. The critical challenges facing energy infrastructure include the following:

1. Ageing infrastructure: Many existing energy infrastructure systems are ageing and need modernization, which can pose challenges for maintenance and reliability.
2. Financing and investment: There is a need for significant investment in new energy infrastructure, which can be challenging to secure in a competitive and changing market.
3. Energy transition: The shift towards renewable energy sources and a more decentralized energy system is putting pressure on traditional energy infrastructure and requiring new energy management and distribution approaches.
4. Climate change: Climate change is driving the need for new infrastructure to support low-carbon energy sources and adapt to the impacts of extreme weather and rising sea levels.
5. Cybersecurity: As energy infrastructure becomes increasingly digitized and connected, it is vulnerable to cyber threats, which pose a significant risk to the reliability and security of the energy system.
6. Public acceptability: Energy infrastructure projects can face opposition from local communities, challenging their development and impacting their long-term viability.
7. Regulatory challenges: Complex and often conflicting regulations can make it difficult to develop new energy infrastructure and slow the implementation of new technologies and innovations.
8. Integration of new technologies: Integrating new technologies, such as electric vehicles, battery storage, and microgrids, into the existing energy infrastructure can present technical and operational challenges.
9. Workforce development: The energy sector is facing a skills gap, with a need for a new generation of workers trained in the latest technologies and practices.
10. Energy access and affordability: Providing access to reliable and affordable energy remains challenging, particularly in rural and remote areas and developing countries.
11. Environmental and social sustainability: Energy infrastructure development must balance economic, environmental, and social considerations to ensure a sustainable energy future.
12. Competition for resources: The energy sector is facing increased competition for limited resources, including land, water, and materials, which can impact the development and operation of energy infrastructure.
13. Interconnections and grid stability: Integrating renewable energy sources into the grid can present challenges in maintaining grid stability and ensuring reliable energy delivery.
14. Energy storage: The development of efficient and cost-effective energy storage solutions is critical to enabling the integration of renewable energy sources into the energy mix.

15. Resilience and disaster preparedness: Energy infrastructure must be designed and operated to withstand natural disasters and other emergencies and ensure quick recovery during disruptions.
16. Cross-border cooperation: Energy infrastructure often crosses national borders, requiring collaboration and coordination between multiple countries to ensure efficient and reliable energy delivery.
17. Data management and analysis: The increasing amount of data generated by the energy sector requires advanced data management and analysis tools to optimize energy infrastructure performance and support decision-making.
18. Public–private partnerships: The development and operation of energy infrastructure often require partnerships between the government and the private sector, which can bring their own challenges and require effective communication and collaboration.
19. Technological advancements: The energy sector is rapidly evolving, with new technologies and innovations emerging all the time. Keeping pace with these developments and ensuring that energy infrastructure can support new technologies can be challenging.
20. Fuel sourcing and transport: The transportation and delivery of fuels, such as natural gas and oil, can present significant logistical and infrastructure challenges, particularly in remote and hard-to-reach areas.
21. Emissions reduction: Reducing greenhouse gas emissions and mitigating the impacts of climate change is a major challenge for the energy sector, requiring a transition to low-carbon energy sources and the implementation of new technologies and practices.
22. Water usage: Energy production and operations often consume significant amounts of water, which can impact water resources and raise concerns about sustainability, particularly in regions with limited water availability.
23. Public safety: Ensuring the safety of the public and the environment is a key concern for energy infrastructure development and operation and requires effective risk management and emergency response planning.
24. Cost-effectiveness: The development and operation of energy infrastructure require significant investment, and it is important to ensure that these investments are cost-effective and provide value for money.
25. Community engagement: Engaging with and addressing the concerns of local communities is critical to the success of energy infrastructure projects and can impact their social and environmental sustainability.
26. Transparency and accountability: The energy sector is subject to close scrutiny, and it is important to ensure that energy infrastructure development and operations are transparent and accountable and that they promote responsible and sustainable practices.
27. Market competition: Competition in the energy market can impact the development and operation of energy infrastructure and can drive innovation and progress in the sector.

28. Energy efficiency: Improving energy efficiency and reducing waste is a critical challenge for the energy sector and can impact the cost and sustainability of energy infrastructure.

29. Labour and safety standards: Ensuring that labour and safety standards are met in the development and operation of energy infrastructure is essential for the well-being of workers and the public and can impact the sustainability and viability of energy infrastructure projects.

30. Long-term planning: Energy infrastructure development requires long-term planning and investment, and it is essential to consider the potential impact of future results, such as technological advancements and changing energy demands when planning for the future.

31. Maintenance and upgrades: Regular maintenance and upgrades are essential for the ongoing performance and reliability of energy infrastructure and can impact its sustainability and cost-effectiveness over the long term.

32. Environmental protection: Energy infrastructure development and operation can impact the environment, and it is crucial to minimize this impact and protect natural resources, wildlife, and ecosystems.

33. Resource allocation: The allocation of resources, such as funding, personnel, and materials, is critical for the success of energy infrastructure projects and can impact their cost-effectiveness and efficiency.

34. Energy security: Ensuring the security of energy supplies and protecting against disruptions is a key challenge for the energy sector and requires effective risk management and contingency planning.

35. International cooperation: Energy infrastructure often transcends national borders, and international cooperation is necessary to ensure its effective development, operation, and maintenance.

36. Cybersecurity: Energy infrastructure is vulnerable to cyber-attacks, and it is important to protect against these threats to maintain its reliability and stability.

37. Fuel diversification: Diversifying the energy mix and reducing reliance on a single fuel source is a challenge for the energy sector and can impact its stability and security.

38. Market dynamics: Fluctuations in energy demand and prices can impact the development and operation of energy infrastructure, and it is important to consider these market dynamics when planning and investing in the sector.

39. Technological integration: Integrating new technologies into existing energy infrastructure can be challenging, and it is important to consider the compatibility and interoperability of these technologies.

40. Consumer education and awareness: Educating and raising awareness among consumers about energy usage and the role of energy infrastructure is vital for promoting responsible energy consumption and for supporting the transition to a more sustainable energy future.

41. Government regulation and policy: Government regulation and policy can impact the development and operation of energy infrastructure, and it is important for the sector to stay abreast of changes and comply with relevant laws and regulations.

42. Social and economic equity: Ensuring that the benefits of energy infrastructure are distributed equitably among different social and economic groups is important for promoting sustainable and inclusive economic growth.
43. Access to finance: Access to finance is a critical challenge for the energy sector, and it is essential to ensure that energy infrastructure projects have adequate funding to support their development and operation.
44. Community health and well-being: The development and operation of energy infrastructure can impact the health and well-being of local communities, and it is vital to assess and mitigate these impacts to ensure their sustainability.
45. Renewable energy integration: Integrating renewable energy sources into the energy mix is a critical challenge for the energy sector, and it requires adequate planning, investment, and coordination to ensure its success.

9.3   How to Deal with All Energy Infrastructure Challenges

Dealing with the challenges facing the energy infrastructure sector requires a multi-faceted approach, incorporating the following strategies:

1. Invest in research and development: Investing in research and development is critical for advancing the energy sector and addressing the challenges facing energy infrastructure.
2. Implement best practices: Adopting and implementing best practices in the development and operation of energy infrastructure can help to improve performance and address challenges.
3. Foster innovation: Encouraging innovation and experimentation in the energy sector can drive progress and help to address the challenges facing energy infrastructure.
4. Promote collaboration: Collaborating with stakeholders, including governments, industry, and civil society, can help to address the challenges facing energy infrastructure and promote its sustainable development.
5. Encourage sustainability: Promoting sustainable energy practices and reducing the environmental impact of energy infrastructure can help address the sector's challenges.
6. Increase transparency and accountability: Increasing transparency and accountability in the energy sector can help to address the challenges of cost-effectiveness, community engagement, and market competition.
7. Implement effective regulation and policy: Implementing effective law and policy can help to address the challenges of energy security, technological integration, and social and economic equity.
8. Provide education and training: Providing education and training to industry professionals, workers, and consumers can help to address the challenges of labour and safety standards, consumer awareness, and renewable energy integration.

9. Foster international cooperation: Foster international cooperation and collaboration to address the challenges of cross-border energy infrastructure and global energy security.

10. Promote renewable energy: Promoting renewable energy sources and reducing reliance on fossil fuels can help address the challenges of fuel diversification, carbon emissions, and environmental protection.

11. Enhance energy efficiency: Enhancing energy efficiency can help to address the challenges of energy demand and usage, reduce costs and emissions, and improve energy security.

12. Emphasize energy storage: Developing and promoting energy storage solutions can help to address the challenges of intermittency and variability of renewable energy sources.

13. Develop intelligent grid technology: Developing and implementing innovative grid technology can help to address the challenges of grid reliability, energy security, and renewable energy integration.

14. Promote electric vehicles: Promoting electric vehicles and expanding charging infrastructure can help address fuel diversification challenges and reduce carbon emissions.

15. Support community engagement: Engaging and empowering local communities in the development and operation of energy infrastructure can help address social and economic equity challenges and community health and well-being.

16. Promote investment: Encouraging investment in the energy sector, especially in energy infrastructure, can help address financing and resource allocation challenges.

17. Strengthen cybersecurity: Strengthening cybersecurity measures can help to address the challenges of cyber threats and energy security.

18. Monitor and assess progress: Regularly monitoring and evaluating the progress of energy infrastructure projects can help to identify areas for improvement and to address challenges as they arise.

19. Encouraging public–private partnerships can help leverage private-sector investment and expertise to address the challenges of financing, resource allocation, and sustainability challenges.

20. Foster technology transfer: Fostering technology transfer and knowledge sharing between countries can help to address the challenges of technology integration and innovation.

21. Develop risk mitigation strategies: Developing risk mitigation strategies can help to address the challenges of uncertainty, volatility, and risk in the energy sector.

22. Foster stakeholder engagement: Engaging and consulting with all stakeholders, including industry, governments, civil society, and communities, can help to address the challenges of community engagement and social and economic equity.

23. Encourage market competition: Encouraging and fostering competition can help address market structure, pricing, and cost-effectiveness challenges.

24. Promote energy access: Promoting access to energy for all, especially for rural and underserved communities, can help to address the challenges of energy poverty and social and economic equity.
25. Foster a culture of sustainability: Fostering a culture of sustainability and environmental responsibility throughout the energy sector can help address environmental protection and resource depletion challenges.
26. Encourage data-driven decision-making: Encouraging the use of data and analytics in decision-making can help to address the challenges of information asymmetry and improve the efficiency and effectiveness of energy infrastructure development and operation.
27. Promote digitalization: Promoting digitalization and technology use can help address the challenges of efficiency, cost-effectiveness, and renewable energy integration.
28. Encourage stakeholder investment: Encouraging stakeholder investment in energy infrastructure projects can help address financing and resource allocation challenges.
29. Develop standardization and certification: Developing standardization and certification programs can help to address the challenges of quality control and safety standards.
30. Foster cross-sectoral collaboration: Foster collaboration between the energy sector and other sectors, such as transportation, agriculture, and industry, can help to address the challenges of energy-sector interdependence and sustainability.
31. Promote international best practices: Promoting international best practices and benchmarking can help address technological integration, innovation, and sustainability challenges.
32. Enhance infrastructure resilience: Enhancing the resilience of energy infrastructure to natural disasters, and other hazards can help to address the challenges of energy security and sustainability.
33. Foster public awareness: Foster public awareness and education on energy issues can help address consumer awareness and sustainability challenges.
34. Encourage environmental responsibility: Encouraging ecological responsibility and sustainable practices throughout the energy sector can help address environmental protection and resource depletion challenges.
35. Develop alternative financing models: Developing alternative financing models, such as green bonds, crowdfunding, and impact investing, can help to address the challenges of financing and resource allocation.
36. Foster innovation: Fostering innovation and research and development in the energy sector can help address technological integration and sustainability challenges.
37. Enhance grid interconnections: Enhancing grid interconnections between countries can help to address the challenges of energy security, reliability, and renewable energy integration.

38. Encourage sustainable procurement: Encouraging sustainable procurement practices, such as purchasing renewable energy and energy-efficient products, can help address sustainability and environmental responsibility challenges.
39. Foster community energy initiatives: Foster community energy initiatives, such as community-owned renewable energy projects, can help to address the challenges of community engagement and social and economic equity.
40. Enhance regulatory frameworks: Enhancing regulatory frameworks and policies can help address market structure, pricing, and cost-effectiveness challenges.
41. Promote sustainable supply chains: Promoting sustainable supply chains, such as using sustainable and environmentally responsible suppliers, can help address environmental protection and resource depletion challenges.
42. Foster public–private collaboration: Collaboration between the public and private sectors can help address financing, resource allocation, and sustainability challenges.

9.4 What is the Main Role of Artificial Intelligence (AI) Utilization in Energy Infrastructure

Artificial Intelligence has a significant role in the energy infrastructure industry. AI algorithms and techniques can be used for a variety of tasks, including:

- Predictive maintenance: AI algorithms can be used to monitor and analyze data from energy infrastructure assets, such as wind turbines and power plants, to predict when they might fail, allowing operators to perform maintenance before issues arise,
- Energy demand forecasting: AI can be used to predict future energy demand, enabling energy companies to better plan and manage their energy generation and distribution systems,
- Energy efficiency: AI algorithms can be used to optimize energy usage in buildings, homes, and industrial facilities, reducing energy waste and lowering energy costs,
- Renewable energy integration: AI can be used to manage the integration of renewable energy sources into existing energy grids, enabling a more efficient and reliable energy infrastructure,
- Smart Grid Management: AI algorithms can be used to manage and optimize energy distribution networks, known as smart grids, which use real-time data to match energy supply with demand,
- Energy Storage: AI can be used to optimize energy storage systems, helping to balance energy supply and demand, reduce waste, and improve the efficiency of energy storage,
- Grid Security: AI algorithms can be used to monitor energy infrastructure for signs of cyberattacks or other security threats, helping to ensure the secure and reliable operation of energy systems,

- Environmental Monitoring: AI can be used to monitor environmental factors that impact energy production and usage, such as weather patterns and air quality, to improve energy planning and management,
- Carbon Capture and Storage: AI can optimize carbon capture and storage technologies, which are used to reduce carbon emissions from power plants and other energy infrastructure,
- Predictive Analytics: AI algorithms can analyze large amounts of energy data to identify patterns and predict future energy trends, helping energy companies make more informed decisions,
- Renewable Energy Optimization: AI can optimize the operation of renewable energy sources, such as wind and solar, to maximize energy production and reduce waste,
- Real-time Monitoring: AI algorithms can monitor energy infrastructure in real-time, providing energy companies with real-time data and alerts to help prevent issues and improve operations, and
- Grid Decentralization: AI can support the decentralized energy grid, where energy is generated and consumed locally, helping to improve energy efficiency and reduce the need for an extensive, centralized energy infrastructure.

AI has the potential to revolutionize the energy industry by enabling new levels of efficiency, reliability, and sustainability. Using AI to automate and optimize energy infrastructure can help meet the growing demand for energy while reducing the environmental impact.

9.5 Conclusion

In conclusion, the energy infrastructure sector's challenges are numerous and complex. Still, these challenges can be effectively addressed with a combination of innovative solutions, cross-sectoral collaboration, and a focus on sustainability. By promoting investment in renewable energy, digitalization, stakeholder engagement, and international best practices, we can build a more resilient, efficient, and sustainable energy infrastructure for the future.

In addition, addressing the challenges facing the energy infrastructure sector requires a long-term commitment and continuous effort from all stakeholders, including industry, governments, civil society, and communities. By fostering public–private partnerships, promoting technological advancements, and enhancing regulatory frameworks, we can build an energy infrastructure that is efficient, cost-effective, environmentally responsible, and socially equitable.

Ultimately, the goal should be to create an energy infrastructure that can meet the world's growing energy needs while also addressing the critical issues of energy security, environmental sustainability, and social and economic equity. Taking a proactive and integrated approach can create a future powered by clean, reliable, and accessible energy.

Here are some additional areas of future study in the energy infrastructure sector:

1. Energy market design and regulation: Research ways to optimize the design of energy markets and regulation to promote sustainable, efficient, and equitable energy infrastructure,
2. Financing and investment: Studies on the financing and investment mechanisms needed to support the development and maintenance of energy infrastructure, particularly in emerging economies,
3. Energy transition and low-carbon development: Research into the challenges and opportunities of transitioning to a low-carbon energy system and the role of energy infrastructure in this process,
4. Rural electrification: Studies on the challenges and best practices for rural electrification, including developing off-grid and microgrid solutions,
5. Energy policy and governance: Research into the role of energy policy and governance in shaping the development and operation of energy infrastructure, including the impact of government regulations, subsidies, and incentives, and
6. Public awareness and engagement: Studies on the role of public awareness and engagement in shaping energy infrastructure development and operation, including the impact of stakeholder engagement, public education, and media outreach.

References

1. Noor, S., Yang, W., Guo, M., van Dam, K.H., Wang, X.: Energy Demand Side Management within micro-grid networks enhanced by blockchain, Appl. Energy. **228**, 1385–1398 (2018). https://doi.org/10.1016/j.apenergy.2018.07.012
2. Ji, C., Wei, Y., Poor, H.V.: Resilience of energy infrastructure and services: modeling, data analytics, and metrics. Proc. IEEE. **105**, 1354–1366 (2017). https://doi.org/10.1109/JPROC.2017.2698262
3. Mignacca, B., Locatelli, G., Velenturf, A.: Modularisation as enabler of circular economy in energy infrastructure. Energy Policy **139**, 111371 (2020). https://doi.org/10.1016/j.enpol.2020.111371
4. Li, H., Yazdi, M.: Advanced decision-making neutrosophic fuzzy evidence-based best–worst method BT—Advanced decision-making methods and applications in system safety and reliability problems. In: Li, H., Yazdi, M. (eds.) Approaches, Case Studies, Multi-criteria Decision-Making, Multi-objective Decision-Making, Fuzzy Risk-Based Models. Springer, Cham, pp. 153–184 (2022). https://doi.org/10.1007/978-3-031-07430-1_9
5. Li, H., Yazdi, M.: Integration of the Bayesian network approach and interval type-2 fuzzy sets for developing sustainable hydrogen storage technology in large metropolitan areas BT—Advanced decision-making methods and applications in system safety and reliability problem. In: Li, H., Yazdi, M. (eds.) Springer, Cham, pp. 69–85 (2022). https://doi.org/10.1007/978-3-031-07430-1_5
6. Li, X., Han, Z., Yazdi, M., Chen, G.: A CRITIC-VIKOR based robust approach to support risk management of subsea pipelines. Appl. Ocean Res. **124**, 103187 (2022). https://doi.org/10.1016/j.apor.2022.103187

7. Li, H., Yazdi, M.: dynamic decision-making trial and evaluation laboratory (DEMATEL): improving safety management system BT—advanced decision-making methods and applications in system safety and reliability problems. In: Li, H., Yazdi, M. (eds.) Approaches, Case Studies, Multi-criteria Decision-Making. Springer, Cham,: pp. 1–14 (2022). https://doi.org/10.1007/978-3-031-07430-1_1

8. Garg, A., Naswa, P., Shukla, P.R.: Energy infrastructure in India: Profile and risks under climate change. Energy Policy **81**, 226–238 (2015). https://doi.org/10.1016/j.enpol.2014.12.007

9. Mignacca, B., Locatelli, G., Velenturf, A.: Modularisation as enabler of circular economy in energy infrastructure. Energy Policy **139**, 111371 (2020). https://doi.org/10.1016/j.enpol.2020.111371

10. OpenAI: ChatGPT [Software] (2021). https://openai.com

11. Sharma, M., Joshi, S., Kumar, A.: Assessing enablers of e-waste management in circular economy using DEMATEL method: an Indian perspective. Environ. Sci. Pollut. Res. **27**, 13325–13338 (2020). https://doi.org/10.1007/s11356-020-07765-w

12. Yazdi, M., Khan, F., Abbassi, R., Rusli, R.: Improved DEMATEL methodology for effective safety management decision-making. Saf. Sci. **127**, 104705 (2020). https://doi.org/10.1016/j.ssci.2020.104705

13. Nedjati, A., Yazdi, M., Abbassi, R.: A sustainable perspective of optimal site selection of giant air—purifiers in large metropolitan areas. Springer, Netherlands (2021). https://doi.org/10.1007/s10668-021-01807-0

14. Jiang, G.-J., Huang, C.-G., Nedjati, A., Yazdi, M.: Discovering the sustainable challenges of biomass energy: a case study of Tehran metropolitan. Environ. Dev. Sustain. (2023). https://doi.org/10.1007/s10668-022-02865-8

15. Li, H., Yazdi, M., Huang, H.-Z., Huang, C.-G., Peng, W., Nedjati, A., Adesina, K.A.: A fuzzy rough copula Bayesian network model for solving complex hospital service quality assessment. Complex Intell. Syst. (2023). https://doi.org/10.1007/s40747-023-01002-w.

16. Jiang, G.-J., Chen, H.-X., Sun, H.-H., Yazdi, M., Nedjati, A., Adesina, K.A.: An improved multi-criteria emergency decision-making method in environmental disasters. Soft Comput. (2021). https://doi.org/10.1007/s00500-021-05826-x

17. Li, H., Guo, J.-Y., Yazdi, M., Nedjati, A., Adesina, K.A.: Supportive emergency decision-making model towards sustainable development with fuzzy expert system. Neural Comput. Appl. **33**, 15619–15637 (2021). https://doi.org/10.1007/s00521-021-06183-4

18. Yazdi, M.: Acquiring and sharing tacit knowledge in failure diagnosis analysis using intuitionistic and Pythagorean assessments. J. Fail. Anal. Prev. **19**, (2019). https://doi.org/10.1007/s11668-019-00599-w

19. Pirbalouti, R.G., Dehkordi, M.K., Mohammadpour, J., Zarei, E., Yazdi, M.: An advanced framework for leakage risk assessment of hydrogen refueling stations using interval-valued spherical fuzzy sets (IV-SFS). Int. J. Hydrogen Energy. (2023). https://doi.org/10.1016/j.ijhydene.2023.03.028.

Printed in the United States
by Baker & Taylor Publisher Services